The Institute of Refrigeration Proceedings 2023-2024

Volume 120

Held multiple dates in 2023 and 2024

ISBN: 979-8-3313-0542-0

Printed from e-media with permission by:

Curran Associates, Inc.
57 Morehouse Lane
Red Hook, NY 12571

Some format issues inherent in the e-media version may also appear in this print version.

Printed with permission by Curran Associates, Inc. (2025)

For permission requests, please contact The Institute of Refrigeration (IOR) at the address below.

The Institute of Refrigeration (IOR)
Kelvin House, 76 Mill Lane
Carshalton, Surrey SM5 2JR
Charitable Incorporated Organisation 1166869

Phone: +44 (0) 20 8647 7033

www.ior.org.uk

Additional copies of this publication are available from:

Curran Associates, Inc.
57 Morehouse Lane
Red Hook, NY 12571 USA
Phone: 845-758-0400
Fax: 845-758-2633
Email: curran@proceedings.com
Web: www.proceedings.com

IOR.org.uk
REFRIGERATION AIR CONDITIONING HEAT PUMPS

The knowledge hub for refrigeration, air conditioning and heat pumps

The Institute of Refrigeration

PROCEEDINGS 2023–24
Volume 120

The Institute of Refrigeration (IOR) is an independent registered charity run for the public benefit. Its membership is over 2200 individuals whose work depends on refrigeration, air conditioning and heat pumps, providing technical information for designers, specifiers, installers, end users and service and maintenance personnel.

IOR publishes a range of Refrigerant Codes of Practice, Guidance Notes, Good Practice Guides and Safety Alerts. The ever growing library of up to date and reference publications contributes to making the IOR a central point of reference and resource for people for all matters relating to refrigeration, air conditioning and heat pump practice. IOR also offers events, webinars and e-learning. Visit www.ior.org.uk/membership for further information about our activities and how you can get involved.

Published by

Institute of Refrigeration
Kelvin House, 76 Mill Lane
Carshalton, Surrey SM5 2JR
Charitable incorporated organisation 1166869

Tel: +44 (0) 20 8647 7033

ior@ior.org.uk
www.ior.org.uk

The Institute of Refrigeration

PROCEEDINGS
Volume 120
2023-24

CONTENTS

Minutes of Meetings of Members 2023-2024

2nd November 2023
A meeting of 54 members and guests of the Institute of Refrigeration was held at Park Crescent Conference Centre in London to hear Alexander Cohr Pachai FInstR, winner of the J&E Hall Gold Medal, present the paper, "Challenges of, and Options for, Refrigerant Selection". The meeting followed the IOR AGM and was also broadcast using Go To Webinar. It was chaired by Dermot Cotter FInstR, Chair of the Papers and Publications Committee. Questions were asked by Nick Rivers MInstR and Dermot Cotter FInstR. The meeting closed at 5.17pm.

14th December 2023
The December meeting took place as a virtual event and was broadcast using Go to Webinar.
15 members and guests join the live broadcast to hear the paper entitled "The History of the Food Refrigeration and Process Engineering Research Centre, and its Role in Food Refrigeration Research". The meeting was chaired by Dermot Cotter FInstR, Chair of the IOR Papers and Publications Committee. The paper authors Stephen James and Christian James gave the presentation. Following the presentation there was a discussion between the presenters and chair about the importance of the research conducted at the Food Refrigeration and Process Engineering Research Centre. The meeting closed at 5.06pm

8th February 2024
The February meeting took place as a virtual event and was broadcast using Go to Webinar. The paper presentation "Waste Heat from the London Underground – Heating and Cooling" was given by **Dr Henrique Lagoeiro,** winner of the Ted Perry Award. 55 people joined the webinar, and the broadcast was chaired by Dermot Cotter FInstR, Chair of the IOR Papers and Publications Committee. The paper presentation was followed by a discussion between **Dr Henrique Lagoeiro** and Dermot Cotter FInstR with additional questions from Lawrence Leask MInstR. The webinar ended at 5.05pm.

4th March 2024
The March meeting took place as a virtual event and was broadcast using Go to Webinar. 41 members and guests join the live broadcast to hear two papers presented. The first speaker was Chris Wave AMInstR, and his presentation was entitled "The Challenges of Heat Pumps for Domestic Heating." The second paper presented was by Dylan Betts and was a case study presentation "Heat Networks in Rural Areas- The Swaffham Prior Project." The meeting was chaired by Dermot Cotter FInstR, Chair of the IOR Papers and Publications Committee. Questions and comments were submitted by David Butler FInstR, Issa Chaer FInstR, Alan Foster, Graeme Fox FInstR, John Minard MInstR and, William Thompson FInstR. The meeting ended at 5.10pm

11th April 2024
The April presentation took place as a virtual meeting and was broadcast using Go to Webinar. 34 members and guests joined the webinar to hear the paper presentation, "Emissions Assessment of Refrigerated Food Transport in the UK" given by Xinfang Wang of

Birmingham University. The meeting was chaired by Lisa Waters, the events manager of the Institute of Refrigeration.

Questions and comments were submitted by Kwong Fat Chang MInstR, Ben Fielden, Ken Gleeson, Norman Highnam MInstR, Gabriele Hoffmann, Chris Rhodes MInstR, Tom Southall, Tim Thurnham MInstR, Daniel Colbourne FInstR. The meeting ended at 4.58pm.

ANNUAL REVIEW 2022-3

1.1 Objectives and activities

The Trustees have reviewed their objectives and evaluated activities against them in line with the Charity Commission guidance and confirm that the IOR continues to provide public benefit through:

- Activities to address the need to reduce emissions from refrigeration, air conditioning and heat pump technologies including technical events and publications that explore opportunities and challenges and widening our audience participation in such events.
- Promoting apprenticeship takeup and employer training, investigating how we can support training providers, and liaising with the Institute for Apprenticeships and Technical Education and City & Guilds to review current provision. We also encourage and promote diversity within the sector.
- Enhancing professionalism of individuals by encouraging Engineering Council professional registration, issuing Skills Specifications for various roles, offering CPD (Continued Professional Development) opportunities, promoting a Code of Professional Conduct as well as activities to promote equality and diversity.

The Trustees review progress towards achieving these aims at Board meetings. Members participated in revising priorities during 2023 which are now:

1. Leadership in education – supporting apprenticeships and their delivery.
2. Promoting our industry to young people – working with STEMAZING and providing toolkits and web resources for their use.
3. Raising awareness of Heat Pump opportunities, skills and technologies.
4. Emphasising the importance of good service and maintenance to achieve efficiency.

The Impact and success in progressing priorities is reviewed at quarterly Trustee meetings where activity adjusted accordingly. In addition, Trustees monitor membership take up and profile, Engineering Council professional registration levels, opportunities for international and national collaboration by working with other organisations. Financial and resources reports are presented quarterly by the Treasurer.

1.2 Signification activities and results

Technical publications

- A Guidance Notes on Oil Draining, Pressure Relief Valves and Quality Assurance.
- Technical papers on Waste heat in energy systems, Ammonia compression history, Heat pumps and the net zero agenda, Heat Recovery, and Design maintenance efficiency.
- Service Engineer Guides published or updated on Brazing good practice, preventing water leaks, water cooled and air- cooled condensers, compressor types, TEVs, Superheat and Subcooling.
- Conference Papers on low carbon RACHP design, sustainability and collaborative practices.
- Annual Proceedings Journal was published.

International profile

- Participation in an ASHRAE international Associations Alliance and International Institute of Ammonia Refrigeration meetings.
- Working with partners in Europe to promote REAL alternatives training and assessments.
- Supporting membership of the International Institute of Refrigeration.
- An international conference on Sustainability and the Cold chain was held.

Networking and exchange of information

- Free technical webinars on energy efficiency calculations and refrigerant legislation.
- The Women in RACHP network continued to promote diversity with a podcast and training days.

- SIRACH webinars on research into heating and cooling.
- Events were promoted from the IIR, Cold Chain Federation, Cryogenics Society, Institute of Refrigeration Ireland, ASHRAE, Refcold India, South West Society, Hampshire Society.
- Regular webinars on the benefits of membership, becoming a STEM ambassador, Engineering Council registration.
- Regular participation in ACRIB (Air Conditioning and Refrigeration Industry Board) meetings allowed IOR to monitor industry skills initiatives such as ACRIB FGas Skillcard and RACHP UK Skills events, to share technical information and monitor legislation and policy in areas including FGas Regulation, EcoDesign, Building Safety Act and Building Regulations review.

Education and Training

- Funding was made available for three members to take part in STEMAZING training so that they could deliver science activities to school groups.
- STEM Toolkits were developed to provide refrigeration science activities to share in schools.
- Guidance notes were published on Work Experience, Mechanic Skills and STEM Toolkits.
- Students on apprenticeship schemes were supported with free access to IOR publications.
- Winners of Student competitions were invited to join the IOR Annual Dinner.
- A network of STEM Ambassadors continued to meet to share ideas and resources.
- Meetings were held with the Institute for Apprenticeships and Technical Education to review and update the national Apprenticeship standard.

Encouraging and promoting innovation

- Ted Perry Student Research Award was presented to Henrique Lagoeiro of London South Bank University for his work on reusing waste heat.
- The J&E Hall International Gold Medal for practical contribution to the development of RACHP technology was awarded to Alexander Cohr Pachai for his work to encourage the use of environmentally responsible refrigerants.

The IOR benefits from significant contribution of volunteers through:

- Membership of Committees and Working groups that review technical papers and programmes.
- Speakers sharing their expertise through technical talks at webinars and other events.
- Judging panels for awards and medals.
- Reviewing and grading of membership applications.
- Giving talks to schools through the STEM ambassador and STEMAZING initiatives.
- Representing IOR at key industry and technical events and meetings.

1.3 Achievements and performance

Significant activities generating income for the period included:

- Membership and RACHP EngTech Section fees and subscriptions.
- Conferences, Dinners and other networking events.
- Publications related income.
- Project grants and secretarial support to ACRIB.

Success indicators monitored at Trustee meetings included:

- Membership levels and Engineering Council registrations.
- Attendance levels at events.
- Publications downloads.
- PR, social media and website analytics.

Professional Registration

The IOR is a professional affiliate of the Engineering Council and members can apply to join the Engineering Council register of professional engineers a licencing agreement with CIBSE. The Engineering Council provides valuable benchmarks against which the Trustees evaluate IOR activity and standards in areas such as Codes of Conduct, CPD, appeals procedures, diversity, corporate status, membership profile and marketing. The IOR's licence as a Professional Affiliate of the Engineering Council was audited and renewed for a further five-year period during 2022. Annual meetings are held with CIBSE to review the management of IOR registrants.

1.4 Fundraising

IOR does not engage in public fundraising or grant giving. There is some expenditure related to membership marketing and awards for innovation -some of which are funded by donation by external organisations. IOR's main source of income is individual membership subscription.

1.5 Areas of planned future expenditure

Key activities being undertaken in the next period requiring investment of additional resources are focused on strategic priority areas including:

- Education - purchase of materials for STEM Toolkits, funding of further intake of STEMAZING trainees, development of fantaticfridges.com website with content to support apprentices and T-Level students.
- Environment – activities related to promoting new TICR project (Transport Industrial and Commercial Refrigeration) co-ordinated by LSBU which will be fully funded by DESNZ.

GOVERNANCE & MANAGEMENT

2.1 Legal Status

The IOR Constitution sets out the objects, powers, membership processes, decision making, AGM and voting arrangements, functions and duties of trustees including how trustees are appointed and inducted. The Rules set out the processes around grades of membership, membership fee collection and appointment of President, President Elect, Trustees, Hon Treasurer and Chief Executive. It also outlines the roles of Committees and Working Groups.

Appointment of Trustees - The Board of Trustees of the IOR includes six elected trustees, a President, President-Elect, Past President and Hon Treasurer. One-third of the elected Trustees retire from office each year by rotation. Vacancies are advertised by newsletter and on the website. Nominations must be supported by two members. If the number of nominations exceed the number of vacancies an election is held. New Trustees are given information about their legal responsibilities, terms of reference for Trustees, past meeting minutes, strategy documents and organisational, policy and financial background. They take part in an induction briefing with the President and Chief Executive. The President-Elect is voted in by the Membership one year before the current President is due to stand down. Presidents serve a term of two years.

Organisational structure - The Board of Trustees appoints various Committees and Chairs to carry out the work of the Institute as shown below and approves their Terms of Reference which are reviewed every 2 years. Each Committee includes a representative of the Trustees and is supported by a member of the permanent IOR staff. The IOR employs a permanent secretariat of five staff members led by a Chief Executive. Staff remuneration, including that of key staff, and the benefits packages for all staff, are reviewed annually by a Finance Committee comprised of Trustees in line with average cost of living increases and if appropriate exceptional staff performance.

Decision making policies - Trustees meet quarterly with a quorum of five. They may also make decisions by email resolution. Decisions are confirmed by agreement of a majority. Trustees may delegate certain functions to a committee by determining the terms of reference. They may authorise the formation of Branches. The Finance Committee is made up of a minimum of the Hon Treasurer, President and Immediate Past President and has authority for financial policies, monitoring of investment performance and resourcing including remuneration of staff. Policies are in place covering the reimbursement of Trustee expenses, Declaration of conflicts of interest, Diversity, GDPR Data Management, Risk management, Volunteer management, Code of Conduct and Discipline procedure amongst others. Policies are reviewed and audited on a regular rolling basis by agreement of Trustees.

2.2 Trustees & Elections

The members who served as trustees on 1st April 2022 and throughout the year to 31st March 2023 are shown below. The current President, Graeme Fox FInstR, is due to complete his term as President at the AGM in 2024.

Graeme Fox
President

Mike Creamer
Past President

Nick Rivers
Treasurer

Ian Fisher
Elected Member

John Skelton
Elected Member

Lisa-Jayne Cook
Elected/
Co-opted

Juliet Loiselle
Elected Member

Catarina Marques
Elected Member

John Ellis*
Elected Member

Damian
Wiszniewski
Elected Member

Paul Singh
Elected Member

At the AGM in November 2022 the term of office for the following elected members concluded and they stood down as elected members: Juliet Loiselle and Lisa-Jayne Cook. John Ellis stepped down from the Board in November 2023 and Lisa-Jayne Cook as co-opted to fill this vacancy for one year. Two new members were elected: Damian Wiszniewski and Paul Singh.

2.3 Committees

- Annual Dinner – Juliet Loiselle (Chair), P Arrowsmith, R Hurley, S Buckell, A Gittoes, G Fox.
- Education and Training – J Skelton (Chair), J Austin-Davies, S Buckell, R Mapperley, I Dickinson, D Russell, P Singh, D Hudson, M Noon.
- Finance – N Rivers (Chair), M Creamer, G Fox, J Skelton
- International Refrigeration Committee – A Pearson (Chairman), M Camacho, J Evans, R Lawton, R Low, D Latta, J Miller, L McCusker, T Sekban and R Unsworth. The work of the Committee was supported by the following subscribing organisations: Cambridge Refrigeration Technology, GEA, Howden Compressors, Koura, J&E Hall International, Star Refrigeration.
- Membership – D Wiszniewski (Chair), A Thomas, S Kayes, A Riches, L-J Cook, C Marques, C Girdlestone, R Lamb.
- Papers and Publications – D Cotter (Chair), D Butler, K Glass, N Franzen, C Marques, C Vines, J Thompson, P Wilson, J Whitwell.
- RACHP Engineers' Section & ACHPI Management Panel – R Wright (Chair), I Fisher, D Garcia, P Stacey, J Broughton, G Adams, M Nankivell, G Wright, D Sowden, I Lemmon, M Carannante, B Parker, A Radford, N McGoldrick, G Fox.
- Technical – D Paget (Chair), A Pearson, P Arrowsmith, S Benton, D Colbourne, J Evans, C Green, N Cook, N Atkins, N Roberts, N Rivers, P Cowan, P Holzer.
- Environment – G Maidment (Chair), B Churchyard, C Marques, C Green, LJ Cook, G Fox, T Mitchell, T Southall, R Lamb.
- Apprenticeship Steering Group – R Mapperley (Chair) and employers and trainers.
- Women in RACHP Network – L-J Cook (Chair), K Bullivant, J Caden, K Perry, C Fountain, J Loiselle, C Marques, A Prado, L Pogson, G Weston, S Gill.

2.4 Branch & Society Contacts

- East Anglian Refrigeration Society – C Sealy
- Northern Branch – M Morton
- IOR Scotland – D Watson
- Hampshire Society (affiliated) – J Emm
- Birmingham Society (affiliated) – P Singh
- South West and South Wales Refrigeration Society (affiliated) - D Wiszniewski

2.5 External Representation & Co-operation

- Air Conditioning and Refrigeration Industry Board (IOR Directors who served during this period M Creamer, L-J Cook, J Loiselle, I Fisher)
- ASHRAE Associate Alliance (G Fox, M Rodway)
- British Standards Institution and European and International Standards Review Committees (AB Pearson, D Colbourne, S Benton, D Paget, J Evans, M Rodway and others)
- International Institute of Refrigeration (UK delegate, J Evans)
- SIRACH (G Maidment, J Evans, L Waters)
- Carbon Trust (ad hoc)

2.6 Staff Members

- Miriam Rodway – Chief Executive
- Bawani Sarvendran – Accountant
- Clare Tooth – Membership Secretary
- Lisa Waters - Events Manager
- Edith Meyer – Communications Officer

2.7 Professional advisors

- Auditor – Myrus Smith Chartered Accountants, Norman House, 8 Burnell Road, Sutton SM1 4BW
- Legal advisor – Third Sector Law, 28-30 High Street, Guildford GU1 3EL
- Bankers – CAF Bank, 25 Kings Hill Ave, Kings Hill, West Malling, Kent ME19 4JQ

FINANCE & INVESTMENT POLICIES

The Trustees have prepared this report and results in the form of a Statement of Financial Activities in order to comply with the Statement of Recommended Practice for Charities.

The Trustees confirm that they have referred to the guidance contained in the Charity Commission's general guidance on public benefit when reviewing the Institute's aims and objectives and in planning future activities. The range of publications, events and projects provide benefit to the public by improving the effectiveness and efficiency of the refrigeration and air conditioning applications which underpin many aspects of public life. Activities and outputs are made widely available through a well visited website and promoted via trade press, newsletters, social media, and free webinar.

3.1 Financial performance

The Charity Management Accounts budgeted for a small surplus this year as activity levels increased following recovery from the pandemic. Several activities were agreed that would employ funds surplus to immediate requirements in strategic objectives to further the charity's benefit to the public in areas of education specifically promotion of careers and support for apprenticeships. During the year the Institute held three Annual Dinners (two in London and one in Scotland), an international conference and a national annual conference that led to increased expenditure and successfully raised additional funds. Membership levels remained fairly static. A new project (TICR) was entered into and is expected to provide some income over the next three financial years. The IOR maintains a diverse range of income sources including membership fees, events income, publication sales, and externally funded projects in order to mitigate risk of over-reliance on any single income source.

The audited accounts for the year show an operating surplus of £26,056 (2022 loss was £8,115) before investment revaluation. Investment values have reduced by £19,802 after last years significant increases of £51,290. Therefore, there is an overall net increase in funds of £6,254.

The reconciliation of funds to be carried forward shows a total of £807,846 (2022 was £801,592) of which £433,195 are held in restricted funds that can only be used for specific purposes and £374,651 in the general fund whose use is unrestricted.

This funding will allow the IOR to continue to invest reserves surplus to immediate requirements in the furtherance of strategic objectives agreed by the membership and provide for the long term grown of the IOR as well as sufficient contingency for unexpected events.

The Trustees are confident that the IOR is well placed to respond flexibly to business changes and to maintain a high level of services to members and the general public through a mix of in-person and web-based services. The Trustees have considered factors that could affect future financial performance which include reduced membership levels due to economic downturns, further drops in investment values, reduced opportunity to collaborate in funded projects and lower levels of support for physical events. The Trustees have in place cautious financial forecasts to ensure financial stability when some of the current income generating projects that the IOR is involved in come to an end in the next 2-3 years and have implemented plans to encourage membership growth as well as continuing to monitor close investment portfolio performance.

3.2 Management of funds

Reserves Policy - The Trustees have considered and reapproved a prudent reserves policy to minimise the risk to the Charity of unexpected economic events and allow for the development of new areas of activity. This level of reserves is reviewed annually and Trustees confirm that the reserves target should continue to be set at six months operating expenditure, which is currently £214,822. The actual level of reserves available as at 31st March were £312,617 (unrestricted funds of £374,651 less tangible fixed assets of £62,034). The analysis of net assets between funds (Note 18) shows that £78,062 of these funds however, have been designated for use by IOR Branches, the REAL Alternatives Consortium, SIRACH and IRC. The current reserves level is therefore broadly in line with the six months expenditure maximum.

The Trustees have judged to have sufficient funds to continue in operation for at least the next 12 months from the date of confirmation of the accounts, due to the high level of continued activity and engagement by members, the continued membership income levels, and availability of reserves upon which to draw should the need arise. The IOR budget forecasts are for a small surplus for the next two years.

Cash management - Cash surplus to immediate needs is invested prudently in line with recognised good practice for Charities, in a balanced mix of bank and building society deposit accounts CAF Bank (Charity Aid Foundation), Virgin Money Building Society, Cambridge and Counties Bank as well as in tax-free Charity investment bonds, which provide long term growth as well as substantial regular dividends. The IOR maintains a broad range of investments to minimise risk. Over the longer-term investment values have increased beyond that which would have been expected from simple bank-account based savings and these bonds continue to yield dividends. Trustees monitor the value of the investments through reports at each Board meeting.

3.3 Risk management

Trustees carry out a regular review of risks associated with the on-going management of the Charity and a risk register is held. Trustees are satisfied that processes and procedures are in place to mitigate the impact of such risks on the operations of the charity. The principal risks identified have been:

- Staffing – the Board review annually staff welfare, staffing requirements and opportunities to invest in IT to ensure adequate resources are available and that these are deployed efficiently with a business continuity plan to mitigate staffing related risks.
- Office working arrangements – efficient use is made of software and hardware to allow flexible working patterns, remote working, team working with volunteers in different locations, and integration of financial processes. Cloud based computing was introduced during this period to facilitate remote working and mitigate the risk of unavailability of access to files stored on a physical server device in the office.
- Cybersecurity and data protection – IOR as a data owner keeps an up to date GDPR risk

assessment, has processes in place to train staff, manage data sharing with processors, manage any potential breaches. It uses external consultants to ensure cybersecurity, updating of firewalls, prevention of security breaches etc.

- Membership profile – the need to attract more members to ensure a continued membership base in the long-term future is being addressed by improving relationships with training providers and providing networking and profile-raising opportunities for younger people. Activities are also targeting a broader range of refrigeration applications including air conditioning, heat pumps and heating to ensure a balanced representation of those working in the sector and contributors to technical output.
- Over-reliance on a small number of volunteers and balance of member input – the increasing technical and educational output needs an increased volunteer input with a more diverse range of experience. An annual recruitment drive to attract new member volunteers, specifying air conditioning or training experience, has been implemented. The IOR equality and diversity policy encourages equal opportunities and a balance of representation.
- Unpredictability of investment income and market revaluation of investments – remains a risk which the IOR addresses by ensuring a varied portfolio of charity investments, monitoring throughout the year their growth and returns and having a split of surplus funds between investments and bank/building society deposit accounts.
- Compliance with legal and professional registration requirements is monitored through subscription to Charity Commission newsletter guidance, Charity Management resources, involvement in Engineering Council meetings and advice from professional auditor services.
- Business disruption due to COVID-19 pandemic is no longer considered a significant risk.

3.4 Designated and Restricted funds

Expenditure is made directly for the purposes for which these funds were established as shown in the notes to the accounts. Planned expenditure for the next period is as follows:

- Distribution of prize monies from the Ted Perry Fund.
- Presidents' fund income will continue to support education initiatives approved by the Board.
- Mike Tinsley fund is available to support studies by a disabled engineer or technician.
- A Designated Fund for REAL Alternatives is holding funds raised by the project to cover costs of maintaining the project website that hosts free elearning materials.
- A Designated Fund for IRC will allow IOR to maintain international institute membership.
- A Designated fund for SIRACH project will allow continued SIRAC events when external funding is no longer available.
- IOR Branches hold designated funds for local activity programmes.

The financial statements that follow show the current values of the restricted, unrestricted and cash funds of the Institute.

STATEMENT OF TRUSTEES' RESPONSIBILITIES

The trustees are responsible for preparing this Annual Report and the financial statements in accordance with applicable law and United Kingdom Accounting Standards (United Kingdom Generally Accepted Accounting Practice).

The law applicable to charities in England & Wales requires the trustees to prepare financial statements for each financial year which give a true and fair view of the state of affairs of the charity and of the incoming resources and application of resources of the charity for that period. In preparing these financial statements, the trustees are required to:

- select suitable accounting policies and then apply them consistently;
- observe the methods and principles in the Charities SORP 2019 (FRS 102);
- make judgements and estimates that are reasonable and prudent;
- state whether applicable accounting standards have been followed, subject to any material departures disclosed and explained in the financial statements;
- prepare the financial statements on the going concern basis unless it is inappropriate to presume that the charity will continue in operation.

The trustees are responsible for keeping proper accounting records that disclose with reasonable accuracy at any time the financial position of the charity and enable them to ensure that the financial statements comply with the Charities Act 2011, the Charities (Accounts and Reports) Regulations 2008 and the provisions of the trust deed. They are also responsible for safeguarding the assets of the charity and hence for taking reasonable steps for the prevention and detection of fraud and other irregularities.

Graeme Fox CEng MCIBSE FInstR

President and Chair

Date 14th September 2023

IOR.org.uk
REFRIGERATION AIR CONDITIONING HEAT PUMPS
The knowledge hub for refrigeration, air conditioning and heat pumps

Challenges of and Options for Refrigerant Selection

Alexander Cohr Pachai FInstR, Ole Reinholdt Nielsen Martin and Pihl Andersen

Winner of the 2023 J & E Hall Gold Medal

Why you should attend

- Improve your knowledge of the future of refrigerant options
- Find out about the relative strengths and weakness of current refrigerants available
- Hear how end users are overcoming these barriers to successfully adopt low GWP alternative refrigerants safely

Abstract

Numerous viewpoints exist regarding suitable refrigerants for specific applications. Frequently, a solitary refrigerant hazard is selected, leading to the disqualification of the refrigerant. Nevertheless, utilising the refrigerant safely remains a viable option. Each refrigerant possesses its own set of strengths and weaknesses, requiring a thorough exploration of these aspects. In this paper, we will explain what it takes to overcome refrigerant weaknesses.

Hydrocarbons are some of the most used refrigerants in the industry and have been in constant use since the beginning of refrigeration because they were used by the petrochemical industry. Here they knew how to handle these fluids in a safe manner. Also, ammonia was adopted by the industry relatively early. A little later also CO_2 was introduced to the markets, however, CO_2 went out of use and the last traced installation was onboard a fishing vessel in 1955. Since then new plants were not installed before in the second half of the 1990s where CO_2 was promoted by amongst others Prof. Gustav Lorentzen from the Norwegian Technical University in Trondheim.

This paper describes how the working fluids can be utilised in a safe ways. The future is naturally occurring substances. Further, CO_2 and NH_3 are part of the human body – every human body release CO_2 in every breath and NH_3 is in our cells where it is a vital part of our organisms. We release about 17 gram of NH_3 per day on average. The systems discussed here are larger systems excluding residential and commercial systems placed in areas with no limitations of people e.g. supermarkets or restaurants.

1. Introduction

In discussions about future refrigerants, it is often claimed that this or that refrigerant has some disadvantages, but are these claims all true? That a refrigerant is flammable does not necessarily mean it is impossible to use, that it should be disqualified. In this paper, we will have a look at what it takes to overcome these minor challenges. It is not as difficult as some would like to point out. All working fluids have their strong and less strong sides, and this is the challenge that needs to be explained.

Flammability, toxicity, scarcity and rare earth products, global warming potential and formation of PFAS when breaking down in the atmosphere – all topics are used in different ways both as barriers and for scaring off their use. However, following the standards available will provide guidance through most issues where bans imposed by the surrounding reality, are the real limit that will affect the use of a given working fluid – also called refrigerants.

In this paper different barriers will be looked at and discussed – and solutions will be pointed out. Some prefer one solution over others and that is also OK, just know what you are doing and follow the law.

Currently, there is a significant focus on using non-fluorinated working fluids, which has been the preferred solution for many years up until the signature of the first Montreal Protocol agreement in 1987. The first focus was on phasing out the chlorinated fluids because they had a harmful effect on the ozone layer that protects life on Earth from the harmful ultraviolet (UV) light from the sun. When chlorofluorocarbons (CFC) and hydrochlorofluorocarbons (HCFC) became banned, the chemical industry quickly introduced new series of working fluids called hydrofluorocarbons (HFC). However, these fluids still had one disadvantage that also the earlier CFCs/HCFCs also had – a high global warming potential (GWP). Consequently the industry is now being forced into a new regime where the legislative force is on the GWP. At the same time, we now see a big push from another side effect of the breakdown in the atmosphere – the formation of long-lived per- and polyfluoroalkyls (PFAS). The issue here is that when the substances break down in the biosphere the break down will form some of the most stable molecules with no known natural breakdown path and with an expected life of not under 4999+ years. These substances are also very temperature resistant and can resist temperatures up to about 1300°C. Incinerators very often only burn waste at about 800°C, so when the smoke leaves the chimney the PFAS is effectively sent into the sky and when it comes down with the rain it does not discriminate between ecologically driven fields or conventional fields and beast. Humans all over the world already have smaller and larger amounts of these substances in our bodies. This is currently very much a topic that scares many citizens globally. The European Chemical Agency (ECHA) has opened a case in which 5 countries have asked for regulation of at least the most common of the 10.000+ known substances in question – among these also the latest un-saturated HFC substances named HFO in some standards like ISO 817 which is also used by other standards as reference.

The alternatives to fluorinated substances are very well known and described in many ways, often referred to as natural refrigerants. These substances are ammonia (NH_3), carbon dioxide (CO_2), water (H_2O), the family of hydrocarbons and finally air. NH_3 is one of the three most used fluids globally. The other two are R-22, a HCFC, and R-134a, a HFC. It is often discussed in the industry if these natural refrigerants are truly natural – because they can also be produced by industrial processing of natural gas. Therefore, they are also more correctly referred to as naturally occurring substances. In fact, what not everybody is aware of is that the human body uses NH_3 as a breakdown product of amino acids – the human body also uses NH_2 and NH_4 for other processes. Every human on average release about 17 gram of ammonia every day through the kidney. In nature ammonia, apart from being a fertiliser for plants, also serves as a neutraliser for some acids which would otherwise be a problem, ammonia is a strong base.

Currently, we see a competition between the supporters of naturally occurring working fluids to gain market shares. Some parties say one is having the upper hand and some another – truthfully nobody has a complete overview. The market is so vast that there will be more than enough to do for everyone just to get a complete phase-out of all CFC/HCFC and HFC. The applications are many and they are everywhere and the need for refrigeration is growing. It will take years before these fluids are fully phased out of use. The next may well be the unsaturated HFCs because of their decomposition products.

Meanwhile, the proponents of naturally occurring working fluids need to optimise and industrialise future systems but also be aware of the circular economy as not all components and metals will be available forever if the industry does not start to see the benefits of recycling more. Recycling is both good for business and for the environment but also creates new kinds of jobs for those that have less to do in the mining industry.

2. Mastering flammable working fluids

In most countries around the world, it requires technical skills to work with hydrocarbons. The good side of these requirements is that the technology used is not that different from HFC technology. This means that being updated to work on hydrocarbons is relatively easy. The biggest part of skills becoming qualified to work on the pressurized system has been achieved before proceeding to hydrocarbon refrigerants.

When some popular fluorinated refrigerants e.g., R-134a and R-22 are called non-flammable, it is a truth with some modifications. Looking into the ISO 817 it says that they are non-flammable as tested, which also implies that they can be - and they are - flammable under other conditions. This has been seen in the past in both test facilities and chillers in the field. Others are classified as having lower flammability – ISO 817 class A2L, e.g., R-32, are known to be flammable and have caused fatalities in the past. Working with A3, highly flammable according to ISO 817, working fluids should not be that unfamiliar to the trained refrigeration and heat pump technicians. The design is different because you must place ignition sources, like controls and other electric components that are not strictly needed in the safety zone (as defined by ATEX regulations or standards - outside the zone e.g., in a separate cabinet. The safety zone is to be ventilated as described later.

One of the things that make the fluids containing chlorine and/or fluorine is their potential to break down into toxic and corrosive compounds – which cannot happen with hydrocarbons. Only if the concentration becomes high enough both types will deplete oxygen in the air, which is another problem, but with proper ventilation, this issue can be mitigated.

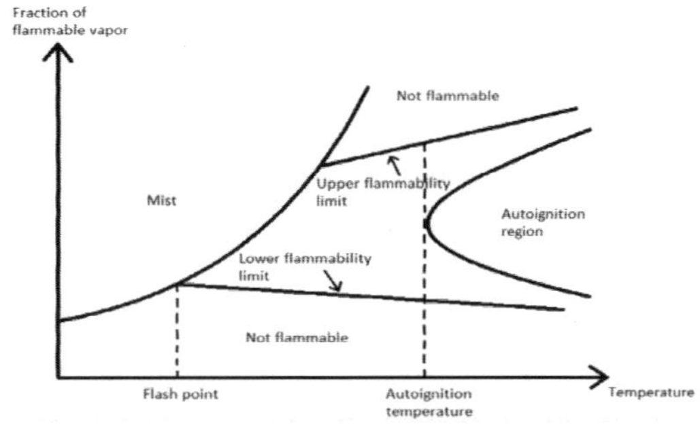

Figure 1 Different regions for flammability of gases

Table 1 Table created from standards, REFPROP , product information (MSDS) and Binbin Yu et al

	Flammabilty limit (% vol)		Flash Point (°C)	Auto inition temperature (°C)	LFL (ppm by vol) ISO 817	LFL (g/m3) ASHRAE sdd. 34
Substance	LFL	UFL				
Ammonia (R-171)	15	28	11	651	167000	116
n-Butane (R-600)	1,6	8,4	-60	420-500	16000	48
iso-butane (600a)	1,8	9,6		462	18000	38
Carbon Monioxide	12	75	-191	600		
Ethane (R-170)	3	12-12,4	-135	515	31000	38
Ethylene/Ethene (R-1150)	2,7	36		490	31000	
Ethylene glycol	3	22	111			
n-heptane	1,05	6,7	-4	204-215		
n-Hexane	1,1	7,5	-22	225-233		
n-Pentane (R-601)	1,5	7,8	-40	260	12000	35
iso-Pentane (R-601a)	1,32	9,16		420	10000	38
Propane (R-290)	2,1	9,5		480	21000	38
Propylene (R-1270)	2	11,1	-108	458	27000	46
HFO-1234yf	6,2	12,3		405	62000	289
HFO-1234ze(E)	6,5-7,6			368	65000	303
HFO-1132a	4,7-5,8	20,3	NA	640	47000	131
HFO-1130(E)					65000	258

Different gases have different flammability properties and Figure 1 shows the principle how it works in general. What you will notice is that the lower flammability limit (LFL) and upper flammability limit (UFL) change with the temperature. The auto ignition temperature has a similar trend. This is especially important to know when these fluids are used in a heat pump application because what can be classified as an A1 or A2L can under other conditions become quite like any other flammable gas, e.g., 1234ze(E) is not flammable under 30°C but at higher temperatures it is flammable. In Table 1 you will see that R-1234ze is flammable under the test conditions used for classification in ISO-817, which is 60°. The information at hand can therefore sometimes seem a little limited and must be treated as such. When considering e.g. which pentane to use in high temperature heat pumps, you can select between R-601 and R-601a. In Table 1 you see that the auto ignition of R-601 is given at 260°C and for R-601a is 420°C. This is a consideration to take into account when e.g., the discharge side can reach temperatures over 160°C. It is therefore required to make proper research and alternatively measurements for the fluids in question at the temperatures required – if you need to know more details. With the fire triangle in mind, it makes it easier to understand the science behind the standards. It is always about removing one of the three items. The fuel concentration can be reduced by ventilation – dilution.

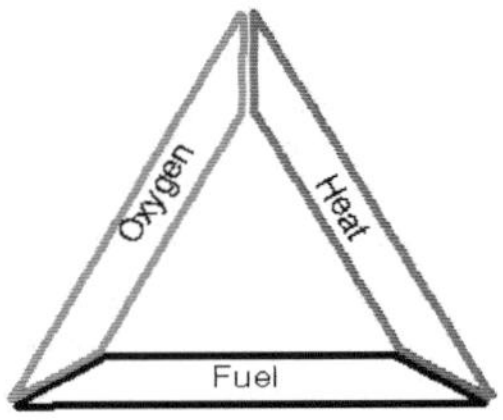

EN 378.2:2016 §6.2.14 say: "The temperature of surfaces that may be exposed to leakage of A2, A2L, B2L, A3, B2, or B3 refrigerants shall not exceed the autoignition temperature of the refrigerant reduced by 100 K." This clause can be especially relevant if you have electrical heaters for heating the machine room during winter conditions and other electrical components, but if ventilated according to EN 60079-10-1, more than 0.5 m/s

say EN 60079-10-1, it becomes less according to Figure 1. You, however, need to do the risk assessment and mitigate the loss of ventilation and/or increased levels of flammable gas. The warmest parts of the system are cylinder heads and discharge pipes.

A sufficiently high ventilation rate (Table 3) is defined as being able to keep the concentration below 50 % of LFL at any given condition. The standards also recommend having all non-essential or not strictly needed ignition sources in the safety zone to be placed out of the zone. This is typically electrical components that can be placed in a separate machine room in a non-hazardous area. Electrical components are to be capsulated to withstand the temperatures needed. Pressure transmitters, solenoid valves or switch valves that are temperature sensitive will have to be removed by extended pipe connections to protect against high temperatures.

EN 60079-0 divides the maximum space/application temperature into different classes for electronic components:

Table 2: EN 60079-0 classification for electronic components

Temperature Class	Maximum surface temperature (°C)
T1	≤ 450
T2	≤ 300
T3	≤ 200
T4	≤ 135
T5	≤ 100
T6	≤ 85

These temperature classes are used for selecting electrical equipment, sensors, transmitters etc.

When it comes to autoignition temperature, small surfaces can have higher temperatures without being a problem, because small objects, e.g., transistors with very small surfaces do not tend to ignite so easily the gas as a larger surface1. However, it is not recommendable to go there if avoidable. The problem with some sensors is the printed circuit board (PCB) in them, which can have a short service life if exposed to very high temperatures. If placed in the air inlet, this can be mitigated.

EN 1127-1 also discuss hot surfaces. Equipment used in areas, e.g., in a special machine room or closed cabinet, in which you can have an explosive gas/air, vapour/air and mist/air atmosphere:

- Category 1: The temperature of all equipment, protective systems and components surfaces which can come into contact with the hazardous atmosphere shall not even in the case of a malfunction – exceed 80 % of the auto ignition temperature of the flammable substance in °C.
- Category 2: The temperature of all equipment which can come in contact with the flammable substance shall not exceed the minimum ignition temperature of the gas or liquid in °C during normal operation and in the case of a malfunction. Where it cannot be excluded that the gas or vapour can be heated to the auto ignition temperature, the surface temperatures shall not exceed 80 % of this temperature measured in °C.
- Category 3: The temperatures of all equipment, protective systems and components surfaces which can come into contact with hazardous atmospheres shall not exceed the auto ignition temperatures of the gas or liquid in normal operation

Following the EN 60079-10-1, you can estimate the leakage rate of your system. Very often you will find that the potential leakage rate is far less than you can even measure with normal detectors or soap water bubbles, which in praxis is about 50 kg/year. Large systems, lose some charge over the years, but following the ventilation required by the standard, you will always find yourself in a non-hazardous atmosphere. In large machine rooms removal of air to cool motors and other equipment can be much more than required for gas removal.

Table 3 Effectiveness of ventilation as given by EN 60079-10-1 (D.1)

Grade of release	Effectiveness of Ventilation						
	High Dilution			Medium Dilution			Low Dilution
	Availability of ventilation						
	Good	Fair	Poor	Good	Fair	Poor	Good, fair or poor
Continuous	Non-hazardous (Zone 0 NE)[a]	Zone 2 (Zone 0 NE)[a]	Zone 1 (Zone 0 NE)[a]	Zone 0	Zone 0 + Zone 2[c]	Zone 0 + Zone 1	Zone 0
Primary	Non-hazardous (Zone 1 NE)[a]	Zone 2 (Zone 1 NE)[a]	Zone 2 (Zone 1 NE)[a]	Zone 1	Zone 1 + Zone 2	Zone 1 + Zone 2	Zone 1 or Zone 0[c]
Secondary[b]	Non-hazardous (Zone 2 NE)[a]	Non-hazardous (Zone 2 NE)[a]	Zone 2	Zone 2	Zone 2	Zone 2	Zone 1 and even Zone 0[d]

[a] Zone 0 NE 1 NE or 2 NE indicates a theoretical zone which would be of negligible extent under normal conditions

[b] The Zone 2 area created by a secondary grade of release may exceed that attributable to a primaryor continuous grade of release: in this case, the greater distance should be taken

[c] Zone 1 is not needed hree i.e. small Zone 0 is int he area where the release is not controlled by the ventilation and larger Zone 2 for when ventilation fails.

[d] Will be Zone 0 if the ventilation is so weak and the release is such that in practice an explosive gas atmosphere exists virtually continuously (i.e. approaching a 'no ventilation condition).

'+' signifies 'surrounded by'
Availability of ventilation in naturally ventilated enclosed spaces is commonly not considered as good.

The standard EN 60079-10-1 in figure (D.1) above gives some good guidance on how to achieve the non-hazardous classification. A heat pump will be a secondary Zone 2 with poor ventilation and a non-hazardous area with good ventilation. Ventilation is key to the chillers and heat pumps based on hydrocarbons (HC). Annex C in the standard provides many good tips on the topic. Secondary is used because the leaks are not part of normal operation and only one leak is expected at a time.

Chillers and heat pumps will be in the secondary grade. In section B.4 of the standard “secondary grade releases are not expected to release in normal operation so, given that it is unlikely that more than one secondary source would release at any one time, only the largest secondary release should be considered”. If you ventilate the machine room efficiently you are having a non-hazardous area. For chillers and heat pumps placed in normal non-hazardous areas, the ambient air can be used to ventilate the machine room. If placed in an ATEX environment the ambient air can be what creates the risk - either a Zone 1 or Zone 0. This situation is very different and difficult to navigate in. Ventilation air from the non-hazardous zone can be far away from where needed. The advice is to follow the guidelines of the site and its risk management team.

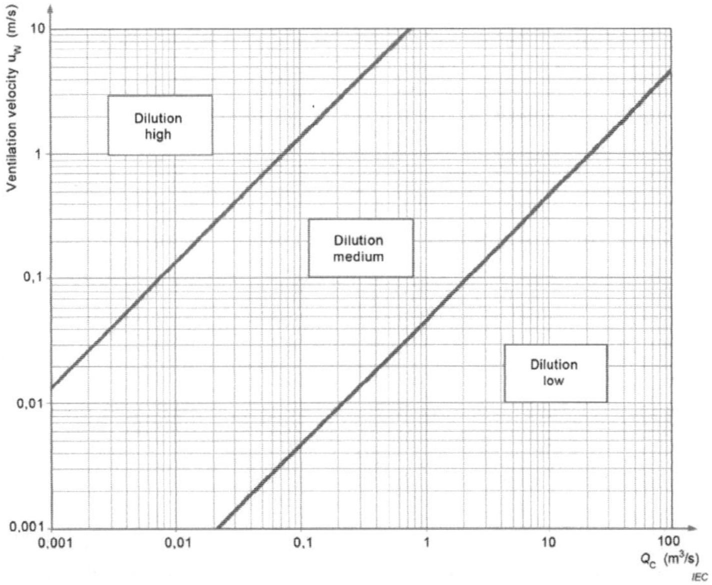

Figure 2 A Chart for assessing the degree of dilution according to EN 60079-10-1 Annex C

Fugitive emissions are small releases of gases or vapours from pressurised equipment due to leaks (generally in an order of magnitude between 10-7 kg/s and 10-9 kg/s, which is about 0,65 kg/year). Though small, these releases can still accumulate in enclosures that are not ventilated. Care must be taken that such emissions are ventilated away so they do no harm, which is not a big deal.

Table 4 Fugitive leaks can be in an order of magnitude between 10-7 kg/s and 10-9 kg/s, here calculated into kg/year

10^-7	
0,00911882	g/s
4792,85161	g/year
4,79285161	kg/year

10^-9	
0,0012341	g/sec
648,64193	g/year
0,64864193	kg/year

IEC 60079-13 emphasises that it is important that the air used for ventilating the space is clean in terms of dust that is flammable. If the air comes from a Zone 2 or higher additional measures must be taken to prevent a situation where dust creates an explosive atmosphere. That includes additional detectors so there is one both inlet and exhaust. The ideal is therefore that the air supplied comes from a place outside any zone – a non-hazardous area. The type, quantity and placement of detectors installed shall be based on a process risk evaluation and the specific standard for gas detection. The location of gas detectors depends on the gas properties and applies to all types of refrigerants.

In flammable atmospheres, the gas detectors shall operate to isolate the electrical power supply of flammable substances if the gas concentration exceeds 25 % of the limiting value.

IEC 60079-13 minimum ventilation flow rate is required to be tested and the test shall be made to verify that the ventilation system can maintain the minimum flow rate required with 50 % of the outlets closed. With flammable working fluids it can also be a solution to build the system into an enclosure and ventilate it to a safe place where you are sure there is no accumulation of gas or potential for gas entering a building air intake. Safety pressure valve outlets should also be vented in the same safe area.

Which standard is the over the over the others as the biggest barrier? EN 378 (ISO 5149) covers some things and IEC standard 60079-10-1 covers the parts where EN 378 is not applicable. You cannot say they stand over one another, but more side by side, supplementing where some things need to be explained better. Otherwise, EN 378 would have to be a very huge document if it had to cover all details of everything. It is very important to remember that local law resides over standards.

Enclosure, cabinet or machine room? Definitions in EN 378 EN 60335-2-24: Particular requirements for refrigerating appliances, ice-cream appliances and ice makers; ".. enclosure, the walls enclosing the appliance as near to all its sides and the top of the appliance as possible, unless the manufacturer indicates in the instructions for installation that a free distance shall be observed from the walls or the ceiling ..". This is the best description of a cabinet we can get.

EN 378-1 §3.2.10 Ventilated enclosure; enclosure containing the refrigeration system that does not enable air to flow from the enclosure to the surrounding space and has a ventilation system that produces airflow from the enclosures to the open air through a ventilation duct.

EN 378-1 §3.2.1 machinery room; enclosed room or space, with mechanical ventilation, sealed from public areas and not accessible to the public, which is intended to contain components of the refrigeration system.

EN 378-1 §3.2.2 separate refrigeration machinery room; machinery room intended to contain only components of the refrigerating system, accessible only to the competent personnel for the purposes of inspection, maintenance, and repair.

One could say that a boxed machine room is a cabinet. This is true when you look in EN 378-2§6.2.15, but this assumes that the main danger of the content comes from electric sparks or components that ignite the gas and therefore EN 60335-2-40 applies.

When you restrict access and remove all controls and place them out of the zone or a safely ventilated cabinet, and reduce electrical components in the zone to an absolute minimum and in a safe construction, e.g. in an ATEX execution, the main danger comes from mechanical vibrations and stress on the pressure system – hence the pressure equipment directive (PED) applies and the EN 378-3§5.14 in which is pointed to EN 60079-10-1 for classification and safety. A machine room includes the entire pressurized system. EN 378-3 §5.14.2.2 is about emergency exhaust ventilation and the requirements for the motors in the ventilation system.

The standards are all connected with references which can make it overwhelming to dig into the different standards. It is therefore important not to be diverted from the actual product/system.

In conclusion, it depends on the application and the safety requirements. If the main danger is electric sparks or components that ignite the gas, then an enclosure is considered a cabinet. If the main danger is mechanical vibrations and stress on the pressure system, then an enclosure is considered a machine room. Since the area is covered by national legislation this must be consulted before considering installation. The law always presides over standards. One should be aware of the fact that some hydrocarbons are more toxic than others or can have other side effects than immediately toxic. Always check the MSDS sheet that comes with the gas in question.

3. Mastering ammonia

Ammonia, NH_3 or R-717, is used globally mostly in larger industrial applications, e.g., abattoirs, dairies, breweries, pharmaceuticals, petrochemical facilities, paper, and wood drying etc. Ammonia is one of the most efficient working fluid for refrigeration and freezing applications[4]. The quality of the ammonia in refrigeration is slightly purer than the ammonia used as fertiliser by farmers worldwide. Ammonia is one of the most produced commodities and the annual global production of ammonia is between 200 and 240 million metric tons[5]. Less than 1 % of this is used for refrigeration, the estimated consumption for refrigeration is between 60.000 and 80.000 tons/year. About 80 % goes as fertiliser, some go to produce explosives and gun ammunition, and some is used for different cleaning products and for colour, e.g., hair colour products.

Many operators perceive ammonia as safe because of the smell and the human reaction when exposed to it and hence, it makes ammonia self-alarming. Only very few individuals cannot smell ammonia but the reactions with watery eyes and coughing are universal. Because the human body knows ammonia it also tolerates ammonia to a certain level around 1000-1500 ppm for 15 minutes without lasting damage[6] . At these levels, nobody will feel comfortable, not even the most hardcore ammonia technician.

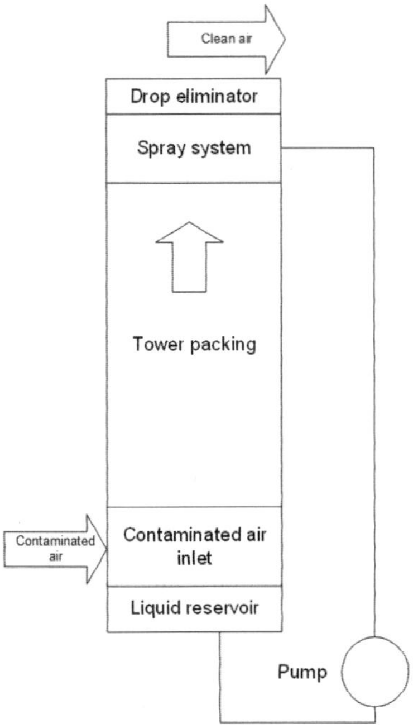

Figure 3 The liquid in the tower is circulated by a pump and the packing insures good exposure of the air to water. To enhance the capacity the water is kept at neutral level (Ph=7) by adding acid to the water.

To avoid emission of ammonia from the machine room you can use a scrubber which can remove the ammonia from the air before it is released to the ambient. This is often used in machine rooms of plants close to rural areas where a major release will cause panic to the public.

A scrubber is a kind of air washing unit. Ammonia vapour is quickly attracted to water when it flows through the shower construction. To increase the capacity of the water, it is possible to maintain a constant neutral pH value by adding acid. However, it will make the water a chemical waste.

Since NH_3 is also a flammable working fluid, a B2L, the ventilation requirements mentioned under hydrocarbons are also applicable here. In many refrigeration and heat pump applications where you use open type compressors and motors, the ventilation requirements are often more related to heat removal than removing the ammonia. Modern and well-serviced ammonia compressors have a negligible leak rate and service technicians rarely have refrigerant cylinders in their service car which is more normal for other markets using other types of fluids. This has to do with generally higher standards such as x-ray of welding's, design of bearings and the system in general. The number of engineering hours in industrial systems are also higher than for commercial systems.

4. Mastering carbon dioxide

Carbon dioxide or CO_2 was one of the early refrigerants and has been around as a working fluid for many years, although it for a period went out of normal use, from about 1955 up to about 1995, where it started to emerge again. Today CO_2 is commonly used both in supermarkets and industrial applications.

Many have asked themselves: CO_2 – but isn't it resulting in the climate change? How can that be a good idea? The simple answers are yes, but as long it is in the refrigeration system, it does no harm to the climate, and should it come out, it has a much lower GWP compared to most alternatives. CO_2 is a by-product from other industrial processes and is just cleaned before it can be used as a refrigerant. GWP of CO_2 is 1 and GWP of flourocarbons from 150 to 4000 or more. Just like ammonia, CO_2 is a natural part of human life, and the body knows CO_2. CO_2 is not toxic but at high levels, the human body reacts to high concentrations and over 10 % of humans will not survive many minutes although there have been examples of recovery after passing out in high concentrations and being there for 30 minutes.

Detectors are available at large in the market and there is no excuse for not installing them to avoid asphyxiation at too high levels. Detectors must have both a visual and acoustic alarm with a sign that says what the alarm is about and what one should do in case it sounds. The same applies for any other working fluid as well.

CO_2 systems have gained a lot of interest since their re-introduction in the middle of the 1990s. In Europe, it has become a kind of standard solution in many applications, especially in supermarkets. Also, heat pumps in district heating are using CO_2 trans-critical heat pumps for temperatures up to about 90°C. The main advantage of CO_2 is that it is compatible with most materials used in the refrigeration industry, however, in the beginning, the higher operating pressures were a limiting factor, but that has long been solved now.

CO_2 in refrigeration grade is available on all continents and in most countries. In countries with beer production CO_2 is recovered from the fermentation and after cleaning it is sold for technical purposes. In countries with volcanic activity e.g., Iceland the CO_2 is taken from the ground and after cleaning it can be used for technical uses. CO_2 can also be recovered from natural gas or from the production of ammonia using natural gas as a source. CO_2 is recovered where the quality and content is high. Burning something to produce CO_2 is very rare because it is not CO_2 rich enough and the cleaning process is more complicated.

By the way, note that carbon dioxide is CO_2 – you sometimes see Co2 – Co is Cobalt

5. Education and vocational training

EN 13313 now in transition to become ISO 22712, is a good and internationally accepted standard with the headline: Refrigerating systems and heat pumps – Competence of personnel. This document defines the activities related to refrigerating systems according to ISO 5149:1 4, and other regional equivalent standards, e.g., EN 378:1 4 and ASHRAE standard 15, the associated competence profiles. It also establishes the competence criteria for persons who carry out these activities. Activities concerning electricity are excluded. It is important to understand that the electrical part is covered by local authorities and other international standards. It is therefore very important to follow these rules as well.

EN 13313/ISO 22712 describes the knowledge subjects and depth of knowledge required for different types of work on a pressurized system. It ranges from the design and pre-assembling of a system over various tasks during the operational life of the system and ends with the dismantling of the system. There is also, as required by the F-gas regulations in Europe, a special requirement for persons with the task of leak checking the system. di

The standard can also serve as a framework for an educational system and training grid for different types of specialists working on the systems, e.g., specialists that focus on leak detection, which is a task and is supported by regulations that focus on leak detection. One must remember the certificates required for each of the competencies.

Recovery of refrigerant is another important task which both from a safety point of view but also from an environmental aspect is important. The competent person must at any time be updated on the changes in regulations both nationally but also internationally.

Enforcement of the rules by the authoritative organ in the local country is responsible for maintaining the competence level at high standards and that the people found in the field working the systems are competent according to the standards.

6. Conclusions

From an end user perspective, the change to other working fluids gives no additional worries, because all standards are clear, and the insurance companies do not consider the risk to be higher than any other heating system if you follow the standards.

The worry can be more about the product and the procedures that have been implemented up to the date of change. As an example, if food production changes from boilers to heat pumps, will that affect the taste or the texture of the final product? If yes, it can become an issue to overcome e.g., for the marketing department. Some years ago, a Norwegian salmon producer changed his equipment and started freezing the salmon filets at -47°C instead of the usual -30°C. The meat texture and taste were perfect, but the colour of the flesh was a little paler than the usual rosa. It was taken up as a challenge for the marketing department.

One barrier that is still needed is the publishing of good case studies. Another barrier is that some case studies are too product and sales oriented instead of focusing on end user experience – especially quotes made by the end user are valuable. Smiling end users are priceless.

The aim of this paper was to demonstrate that the standards are in place and are already being used by many manufacturers in the markets. The capacity of education institutions and the capability to make the new workforce interested in working in the field of refrigeration and heat pump technology is lacking in many countries and regions. It is a pity that only few people are interested in this business because it is very challenging and no day is the same as the previous one.

The challenge of selecting the right refrigerant is not the biggest challenge the industry has, but it takes up more discussion time than it deserves. If help is needed to understand the requirements, help is available but hurry, time is short, and the challenge is big. In this discussion, you could consider delivery risks in case of trade limitations.

Enforcement, enforcement, and enforcement is required to maintain safety and keeping the safety of refrigeration and heat pump systems at the highest possible level.

For further reading about the flammability of working fluids it can be recommended to download the document: (Bulletin 627, Burea o Mines) – Flammability Characteristics of combustible gases and vapors. Author Michael G. Zabetakis. 1965. 7328370 (osti.gov)

Declaration of Competing interests

The authors declare that they have no known competing financial interests or personal relationships that could have appeared to influence the work reported in this paper.

Support

This paper is written in gratitude to J&E Hall for receiving the J&E Hall Gold Medal for innovation, application and advancement of Refrigeration and the prize that followed this award. I wish to thank all involved in the nomination and voting for me to receive this honour, which places me among a select array of highly respected individuals. I would like to thank the reviewers and the team at IoR for assistance I always get when I need it most.

About the authors:

Alexander Cohr Pachai

Alexander Cohr Pachai is an independent consulting engineer with a lifelong experience working with all types of refrigerants from apprentice to senior product specialist and consultant. Mr. Pachai has present papers on many topics within the industry at conferences on several continent. Practical experience has always been the strong side of his activities and approach to things.

Ole Reinholdt Nielsen,

Ole Nielsen holds a master degree in mechanical engineering with specialization in refrigeration. Ole has been involved in Montreal Protocol matters since 1993 in different capacities and since 2011 as Chief of the Montreal Protocol Division at United Nations Industrial Development Organization (UNIDO).

Martin Pihl Andersen

Currently a PhD-student at the Technical University of Denmark, researching in high-temperature heat pumps for industrial applications with a special focus hydrocarbons and steam production. As a part of the Danish, national project SuPrheat, Martin focuses on researching, developing, and demonstrating heat pumps using natural refrigerants where each technology show potential of highest efficiencies.

Reference:

[1] Daniel A. Crowl and Joseph F. Louvar. Chemical Process Safety Fundamentals with Applications. 4th. Pearson, ISBN-13: 978-0-13- 485777-0, ISBN-10: 0134857771, 2019.

[2] Lemmon, E.W., Huber, M.L., McLinden, M.O., 2018. NIST Standard Reference Database 23, NIST Reference Fluid Thermodynamic and Transport Properties. REFPROP version 10.0.

[3] Binbin Yu , Hongsheng Ouyang, Junye SHI, Wuchan LIU, Jiangping CHEN: Evaluation of low-GWP and mildly flammable mixtures as new alternatives for R410A in air-conditioning and heat pump system. https://doi.org/10.1016/j.ijrefrig.2020.09.018

[4] Benefits of Ammonia Use in Refrigeration | Danfoss

[5] https://www.statista.com/statistics/1065865/ammonia-production-capacity-globally/

[6] Toxicological Profile for Ammonia. U.S. Department of Health and human services, Public Health Service, Agency for Toxic substances and Disease Registry, September 2004 ATSDR Ammonia Tox Profile (cdc.gov)

Partial Q&A (due to technical error)- Challenges of and Options for Refrigerant Selection

Nick Franzen – You mentioned air changes being 4 per hour. Where did that figure come from?

Alexander Cohr Pachai – That is a historical level. It is a value that is used but we often see that the ventilation rate is higher and that has to do with the necessity of moving heat out of the motor and compressors in many industrial sites.
You could say one of the differences between the 2 leak scenarios is that one is with a fixed leak so that when you have a leak it doesn't expand, where the other row is where it expands, it rose during the process but I don't see an operation system having a leak where the material erodes away while you are leaking.

Dermot Cotter – A popular option in chillers is Propane and it seems that people are putting them into a nice box. You were talking about ventilation. Would you expect to see with Propane and other chillers in a box that they were ventilated?

Alexander Cohr Pachai – You have to have a leak detector so that you can ventilate in case you get a leak or stop all fans and then you can open the doors and ventilate.

Dermot Cotter – Would you want continuous ventilation?

Alexander Cohr Pachai – Not in a cold room. In a plant room only qualified or competent persons are allowed access.

Dermot Cotter – When it comes to refrigerants like CO2, you talk about hazard and safety, would you prefer to be in somewhere with CO2 or would you prefer to have Ammonia in a plant room?

Alexander Cohr Pachai – This is an interesting point because Ammonia is a kind of silver lining. I don't think anyone would stay too long if you had a major leak into a room of any size whereas CO2 does not have any smell and that is a challenge. We did try ways to make CO2 smell but they were all unworkable so you still have to rely on leak detectors.

Dermot Cotter – The other thing that is becoming popular is putting a fixed gas detector in plant rooms. Do you think that technicians should be wearing personal detectors as well?

Alexander Cohr Pachai – It depends on the plant room. They can be very small or they can be very large and in a large room you can have an accumulation in corners that would not be picked up by the fixed detector.

IOR.org.uk
REFRIGERATION AIR CONDITIONING HEAT PUMPS

The knowledge hub for refrigeration, air conditioning and heat pumps

The History of the Food Refrigeration and Process Engineering Research Centre (FRPERC) and its Role in Food Refrigeration Research

Stephen James and Christian James

Why you should attend

- Gain an understanding of what the FRPERC has been involved with regarding refrigerated food production, storage and retail display over the years
- Review the published outcomes of many of the investigations
- Gain insights into some of the myths as well as the potential for future food refrigeration technologies

Summary

In September 1967, investigations were started on a programme of research into meat refrigeration and processing at the new Meat Research Institute (MRI) under construction at Langford near Bristol. Until July 2009 that programme, now extended to cover the refrigeration, thermal processing, and handling of all foods, continued at the same site under the auspices of the Food Refrigeration and Process Engineering Research Centre (FRPERC). In that time, the MRI having become the Institute of Food Research – Bristol Laboratory (IFR-BL) in 1985 until closure in 1990 and FRPERC having been established in January 1991 at the University of Bristol. This programme of experimental and theoretical studies covered many aspects of the chilling, freezing, thawing, tempering, refrigerated storage, refrigerated transport, refrigerated retail display, domestic handling and refrigerated storage, cooking, pasteurisation and reheating, decontamination, hygienic processing, robotics, automation, and cutting of foods. FRPERC's activities at the University of Bristol ceased in 2009 and at the end of July 2009 some members of the centre moved to the Grimsby Institute to re-establish FRPERC at Grimsby. In April 2023 FRPERC ceased activities at Grimsby. This presentation reviews and celebrates our research carried out between 1967 to 2023 on Food Process Engineering.

These studies have resulted in over 1000 publications, far too many to review in any detail in a few pages. In 2002 we published a book on meat refrigeration [1] covering all the early meat refrigeration studies and in 2009 a review [2] of all our studies between September 1967 to July 2009. In this current review we have therefore outlined some of the more important studies carried out prior to 2009 and reported in more detail on work carried out between 2009 and 2023.

Primary chilling

Chilling is the first and probably the most important process in the cold chain. In 1967 the meat industry required reliable data on the effect of air temperature, velocity, and relative humidity on the cooling rate of and evaporation from beef sides of different weight and fat cover. An Industrial scale refrigeration pilot plant was constructed at Langford and was used to chill over 200 whole beef sides. The comprehensive results of the investigations were presented to the Institute of Refrigeration (IOR) in 1976 [3] and 1977 [4]. For ease of use, the main data were presented as four design charts. Each chart was in the form of a plot of the logarithm of temperature against time at a defined position within the carcass. Another major consideration of industry was weight loss. High accuracy balances (±20 g, over a range from 0 to 250 kg) were used to measure the weight of the beef sides during chilling. Data could therefore be provided on the influence of different chilling and carcass conditions on weight loss. In subsequent years, the same methodology was used to provide similar design charts for goat [5], mutton [6], pig [7], and chicken [8] chilling.

An important result of the studies was that pork could be toughened (cold shortened) during rapid chilling [9]. While a recognised problem with beef and lamb, pork was less prone to cold shortening. But it was show to occur if rapidly chilled. The comprehensive body of work on meat chilling has formed the backbone of meat chiller design over the past 40 years. It provides peer reviewed data sets relating rates of temperature reduction in carcasses over wide weight ranges to the chilling conditions used i.e. air velocity, temperature and RH. It also provides data on the weight loss/yield under different regimes and energy extraction rates required. This was demonstrated in a recent legal dispute in 2022 where experts on both sides relied on the data to support their opinions. Key points in the dispute on the performance of a new carcass chilling system were, 1) was there a testable process design specification, 2) was all the required data on the conditions required available and taken into account when it was designed, 3) could a maximum weight loss be guaranteed, 4) did the operational conditions substantially exceed those in the specified design.

Freezing

Meat has been commercially frozen since the 1880s and our early studies looked at commercial performance and the problems in meeting European Economic Community (EEC) freezing requirements. Later studies at Langford looked at conventional, air, immersion, plate and cryogenic freezing of beef quarters, beef blocks, mutton carcasses, offal, meat products and ready meals.

In the 1990s the need for environmentally friendly food freezing systems focussed our freezing studies on air cycle refrigeration. In operation our first open cycle pilot plant far exceeded expectations and was shown to achieve minimum temperatures of -84°C and hot temperatures of between +40 to +70°C. Further studies resulted in a fluidised bed air cycle freezer and an integrated, rapid heating and cooling system for the food industry.

When moving to Grimsby we inherited a Cells Alive System (CAS) freezing system, a novel freezing system which claimed to improve quality at a time when there was a growing worldwide interest in novel freezing research. CAS claimed to improve food quality by effecting how ice crystals are formed by vibrating water molecules in the food during freezing utilising oscillating magnetic fields. Most if not all the novel freezing studies are based on the belief that freezing is detrimental to the quality of the food being frozen and that the main cause of this detriment is large ice crystals. We had just published a paper comparing rapid and slow freezing of aged beef steaks with unfrozen control samples [10]. In the study trained taste panels found no difference in eating quality between slow, fast, and unfrozen steaks. A small change in the final temperature of the cooked steak had a far larger effect on eating quality than freezing or freezing rate.

While carrying out experimental studies we carried out two comprehensive reviews on novel freezing [11] and dehydrofreezing [12] of food. We found that many of the publications on novel freezing start with the premises that the freezing process is detrimental to food quality, 'fresh being better than frozen', and the quality of frozen is detrimentally affected by the development of large ice crystals in the frozen food. Fast freezing rates are advocated to produce small ice crystals, as do the novel techniques of electo/magnetic or sonic excitation which promote super-cooling and/or limiting nucleation and ice crystal growth. Ours and other reviews, however, provide very little reliable information that taste panel or consumers can detect any difference in the eating quality of cooked previously frozen food that can be related to ice crystal size in the frozen raw material. Further FRPERC studies on electromagnetic excitation during freezing failed to show any quality advantages with garlic [13], pork loin [14], and apples and potatoes [15]. Few of the recent novel freezing publications actually look at eating quality of the product at the point of consumption so there is no real proof that novel systems produce any marked commercial advances.

One exception is some of the developments in dehydrofreezing (partial dehydration of the raw material before freezing). These appear to be able to produce some frozen fruits and vegetables with eating qualities that are very close to fresh products after thawing.

Thawing and Tempering

Interestingly, investigations on meat thawing at Langford preceded those on freezing. Factories that produce pies, pasties, burgers, or other canned, frozen, or chilled meat products require regular supplies of raw material. To take advantage of cheaper raw material arising from seasonal fluctuations in availability and to provide for scarcities, meat is often purchased, frozen and stored until needed. Thawing such material presents difficulties in obtaining consistent reproducible times, bacteriological condition, physical appearance, and loss of weight. Initial studies on pork legs were rapidly expanded to meet other industrial demands for data on thawing systems for beef quarters, lamb and mutton carcasses, meat blocks and cook-freeze catering packs [2]. A key aspect of the thawing studies was the use of finite difference mathematical modelling techniques to augment and expand the range of experimental conditions investigated. Its use became an essential aspect of all the subsequent refrigeration studies [16].

Frozen food is defined as "tempered" when its temperature is raised from that of a solidly frozen condition to some higher temperature, still below the initial freezing point, at which it is still firm but can readily be further processed. In other cases, unfrozen, chilled meat can benefit from "tempering" (partial freezing) to aid processing. At different temperatures in the frozen state, the relative proportions of ice and unfrozen water in food vary. As the percentage of water that is ice decreases, the mechanical properties of the food change and it becomes less hard and brittle. This allows further processing, for example flaking, grinding, or dicing. Our studies showed that conventional air tempering of 25kg meat blocks could take anything between 10s of hours and a week, while an even temperature of between -5 and -2°C could be achieved in 3 to 7 minutes using microwaves at a frequency of approximately 900 MHz [17].

Bacon can have an initial freezing point as low as -11°C due to its high salt content. Optimal high-speed slicing of bacon requires the bacon to be presented to the blade at a very controlled temperature that ensures the correct amount of ice is present to ensure the optimum texture. The temperature depends on the salt and water content and a few °C of error can significantly reduce cutting yield. FRPERC developed a two-stage tempering system for bacon middles which was commissioned by a large retailer who claimed that the

£250,000 cost was recovered in its first year of operation due to increased yield of high-quality slices [18].

Secondary chilling

Meat is chilled immediately after slaughter, ideally to between 0°C and 4°C. Most of the subsequent operations in the cold chain are designed to maintain the temperature of the meat. Cooking is a very common operation in the production of many meat products and operators appreciate the importance of rapidly cooling the cooked product. However, any unit operation such as cutting, mixing, tumbling will add heat to the meat and increase its temperature. A secondary cooling operation is then required to return its temperature to near to 0°C. Over the years, studies were carried out at Langford on many aspects of secondary cooling of large and small individual items, solid/liquid mixtures, and sauces [19].

In the UK the Department of Health Cook-chill Guidelines (1989) recommended maximum cooling regimes and the use of special equipment to reduce product temperatures rapidly after cooking. These guidelines have been widely adopted and are still used across the food industry. Our investigations into air blast cooling of Bolognese meat sauce in metal trays of different depths but the same lateral dimensions showed that, if surface freezing was to be avoided and a simple single stage operation used, only 10 mm depth of product could be chilled within the Guidelines [20].

Chilled and frozen storage

Early Langford storage studies concentrated on improving industrial and commercial profitability by reducing weight loss during chilled and/or frozen storage. The conclusions of our comprehensive reviews of frozen storage life were that much of the underlying data had been gathered from practical experience backed up by a relatively small number of controlled scientific experiments. Much of the scientific data dated back to the time when meat was either stored unwrapped or in wrapping materials that are no longer generally used (such as waxed paper). It was thus not surprising, when consideration was made of the changes in packaging and handling methods over the last century, that there is a considerable scatter in published data on storage lives for similar products.

Much of our work on cold store performance has been confidential, however, Defra have funded two LINK research projects looking at different aspects of cold-store use. In one project a predictive program, 'ColdRoom', was developed to allow cold-storage operators, contractors, and manufacturers to specify and design cold rooms to keep food at optimum temperatures under actual working conditions, and users to rapidly predict the effect of operating conditions and loading patterns on performance and identify how they can avoid unacceptable food temperatures [21].

A separate project looked at air entrainment through cold-store entrances. In cold-stores energy is wasted due to air infiltration into the room during loading and unloading and other instances when the barrier between the cold and warm environments is removed.

In 2018 we were funded by the Meat & Livestock Australia (MLA) to look at the long term frozen storage life of commercially produced lamb and beef. In most of the previous published long term frozen storage life studies, small carefully replicated samples of lean meat, with a known pre freezing history, were stored in laboratory freezers at different temperatures and sampled at set times. In our study commercially produced frozen export lamb and beef (boxed loins and boxed trim) was sourced in Australia. The meat was shipped by sea (beef) or air (lamb) to the UK under commercial conditions and stored in large conventional freezer rooms prior to sampling.

Our aim being to duplicate as carefully as possible a normal cold chain for Australian frozen lamb and beef. There were three overall objectives for this research project: (1) To determine the shelf-life of frozen Australian beef and lamb loin (strip loin and eye of loin, respectively), beef trim of 65CL, 85CL, and 95CL, and lamb trim of 65CL, 85CL, and 90CL (Chemical Lean - a mandatory AUS-MEAT Ltd measurement of the amount of lean red meat compared to the amount of fat in a meat sample);(2) To determine the correlation of frozen end of shelf-life (determined by taste panel) to any measurable parameters, such as oxidative rancidity (PV and TBARS); (3) To determine the rate of oxidation or development of rancidity in frozen Australian beef and lamb meat stored at -12°C, -18°C (Control), and -24°C. After an initial 2 years of storage at either -12, -18 or -24°C the organoleptic and taste panel results showed no marked deterioration, so storage was extended for a further year. Clear changes with time were found in chemical rancidity that could be related to meat composition, packaging, and storage temperature. However, these changes do not appear to correlate clearly with the taste panel results in this study. Whereas the PV and TBARS levels show a clear relationship with storage temperature over time at different holding temperatures, this does not appear to be reflected in the mean taste panel results. While some sensory degradation, small decreases in mean sensory scores, occurred over time in all meat this did not appear to be affected by storage temperature, but occurred at all storage temperatures, and the meat remained palatable after 38 months.

While -18°C has become the standard temperature for the storage of frozen foods, red meat of the type examined appears able to be stored successfully for many months or years at a temperature warmer than this threshold. Providing the meat is of sufficient hygienic quality when frozen and handled under controlled hygienic conditions, no food safety hazards exist with frozen meat that has been held at, or reached, a temperature between -10°C and -18°C. Sensory degradation occurs only slowly at these temperatures and no food safety hazards arise. It is interesting to note that a presentation to the Institute of Refrigeration (IOR) in 1974 [22] reported that UK frozen meat at the time was transported and stored typically at -9 to -10°C, and had been for years. The practices were only changing due to the requirements of regulations for 'intervention freezing' and the increasing use of general food cold stores rather than dedicated meat cold stores. While -18°C has become the standard temperature for the storage of most frozen foods, red meat of the type examined appears to be able to be stored successfully for many months or years at a temperature warmer than this threshold. The results of our study are currently being prepared for peer reviewed publication.

Super-cooling

Super-cooling is an alternative process with the potential to extend shelf-life in comparison with chilling without the damage that happens with super-chilling or freezing. Super-cooling is the physical phenomena where the temperature of a substance can be reduced below its freezing point without ice crystallization occurring. While the mechanism in pure liquids, such as water, is attributed to the lack of a seed crystal or nucleus around which a crystal can form during freezing it is less clear how super-cooling occurs in more complex liquids and solids, such as foods. Research begun at Langford and continued at Grimsby has looked at the super-cooling of liquid and solid foods. Initial studies [23] showed that garlic bulbs and cloves could be super-cooled and stored well below their freezing point (of -2.7°C) at -6°C. Further research by us has looked at super-cooling in fruits, vegetables, meats, fish, and other foods [24] and a current continuing research project is looking at how super-cooling can be implemented in the fresh fish industry.

Although the literature often states that super-cooling is unstable in solid foods, we have found (in unpublished work) that retail packed beef and fish can be held in super-cooled state for at least 35 and 20 days, respectively. Other independent studies have also demonstrated stable super-cooling in other foods. While we have seen stable super-cooling without any assistance there is also some evidence that technologies such as magnetic fields may help maintain super-cooling in different products.

Transportation

Early transportation studies at Langford focused on the problems of poor temperature control during road transportation of both chilled and frozen meat. We found that there are substantial difficulties in maintaining the temperature of chilled foods transported in small, refrigerated vehicles that conduct multi-drop deliveries to retail stores and caterers. The vehicles must carry a wide range of products and operate under diverse ambient conditions. During any one delivery run, the chilled product can be subjected to as many as fifty door openings, where there is heat ingress directly from outside air and from personnel entering to select and remove product. The design of the refrigeration system must allow for extensive variation in load distribution, which is a function of different delivery rounds, days of the week and the removal of product during a delivery run. A refrigeration system's ability to respond to sudden demands for increased refrigeration is often restricted by the power available from the vehicle. All these problems combine to produce a complex interactive system.

A predictive program was developed from individual modules that were verified either using a static vehicle body in our processing halls or on the road. The resulting program, called 'CoolVan' was a unique tool for vehicle operators and manufacturers. Much effort was put into making the program very user friendly and easy to use resulting in very favourable feedback from the many companies involved, and the program was released subsequently as a commercial product [25].

Retail display

The retail display of refrigerated food has long been considered one of the weakest links in the cold chain. Controlling the temperature of an individual retail pack is far more difficult than that of a bulk pack or pallet load of product. In addition, the display cabinet must cater for the marketing requirement for a minimal barrier between the customer and the product. Studies on delicatessen cabinets for unwrapped food showed that colour changes due to surface desiccation was the factor limiting display life and the RH in the cabinet the key controlling parameter [26].

Projects have also looked into the use of environmentally friendly refrigeration systems, especially the use of the air cycle, for retail display. The application of air cycle to retail display resulted in prototype chilled and frozen display cabinets with temperature control, noise and energy levels that were as good as, if not better than, conventional cabinets [27].

While FRPERC and others have produced much data on the performance of retail display cabinets under controlled conditions there is very little data on the temperatures in cabinets in real retail stores, particularly data on performance over weeks of operation. In a recent study we monitored temperatures in frozen retail display cabinets in four branches of a major UK supermarket chain, ranging from a medium size outlet to a large hypermarket [28]. Two types of cabinets were monitored which were: (1) combined frozen top doored and open bottom (well) cabinets (ISO classification RYF3) and (2) vertical frozen glass doored cabinets (ISO classification RVF4). The results showed that while cabinets may run at the required overall temperature(s), some frozen products in those cabinets can spend a substantial amount of time, up to 45 %, above -12°C, at sub-optimal temperatures. While this will have no effect on the safety of the food, it is likely to have a detrimental impact on quality.

Domestic handling

The past decades have seen a considerable increase in legislation defining maximum temperatures during the production, distribution and retailing of chilled food. However, as soon as the food is purchased by the consumer, it is no longer covered by any of these legislative requirements.

After a chilled product is removed from a retail display cabinet it is outside a refrigerated environment whilst carried around the store and transported home for further storage. In the home it may be left in ambient conditions or stored in the refrigerator until required. FRPERC carried out its first survey of the performance of domestic refrigerators in UK homes in the late 1980s [29] and we published our last review on the performance of fridges in 2017 [30]. The overall mean temperature inside a domestic refrigerator has remained at approximately 6°C throughout that period with only a small proportion reliably keeping chilled food within the optimal temperature quality range.

Energy

While many of our projects looked at multiple aspects of the cold-chain at the same time, one particular project encompassed energy usage in the complete cold-chain. In the UK there is increasing pressure on government and industry to make significant reductions in carbon emissions. In the UK 11% of electricity is consumed by the food industry and in some sectors a substantial portion of site energy, up to 90%, is consumed by refrigeration systems. Starting in June 2006, Defra funded a project to "identify, develop and stimulate the development and application of more energy efficient refrigeration technologies and business practices for use throughout the food chain whilst not compromising food safety and quality". This project, led by FRPERC, was a collaboration between 4 top UK University research groups, an industrial steering group, and hundreds of stakeholders from the food and refrigeration industries. The first important findings of this project were presented to the IOR in February 2009 [31]. The research programme concentrated on three topics: (1) Mapping of energy use; (2) Identifying new technologies; and (3) Feasibility studies on promising technologies. In the mapping exercise, we identified and ranked the top 10 food refrigeration operations in terms of the potential to achieve the greatest total reduction in energy usage. We identified large energy savings that can be achieved with current knowledge and the barriers to their uptake. Clear gaps were identified in 'real' energy data and in current knowledge in areas of technology, equipment, management, maintenance, and overall understanding. We critically reviewed all the 'new/alternative' refrigeration technologies in terms of their potential to save energy in a food refrigeration operation and their probable development timescale before they are likely to enter commercial use. Finally, in silico (computer-based) studies resulted in a powerful tool to optimise the energy consumption of a total 'food refrigeration' process. This was used to direct further experimental studies, provide case histories, and identify further research needs.

In 2010 we produced a review on the relationship between the cold chain and climatic change [32]. This and a further book chapter [33] have become some of our most referenced publications. Food refrigeration is a major user of energy throughout the world and any increase in ambient temperatures will increase the energy consumption of refrigeration systems. However, the growing of crops and rearing of animals requires considerably more energy than refrigeration and this is imbedded in the food. If this food is wasted rather than consumed, then the energy loss is substantial. Chilling and freezing food can considerably extend the storage life of all perishable foods i.e. meat, fish, fruits and vegetables, prepared food and grains. Our conclusions were that increasing the amount of food efficiently refrigerated would substantially increase the overall sustainability of worldwide food production even after considering climatic change.

Other research interests

FRPERC were one of the first food engineering groups to look at the use of robots in food processing in the late 1980s. An initial study was carried out on the application of robotics to poultry processing which revealed seven potential areas for the application of robotics in poultry processing, hanging of live birds, evisceration, grading, trussing, portioning, giblet packaging and further processing. Later FRPERC, in conjunction with European project partners, implemented robotic butchery systems to separate a pork carcass into the major sections (known as primals) in Norway and Spain. This resulted in a commercially viable robotic pork primal cutting robot system.

The work on robotic butchery illustrated a lack of fundamental understanding of food cutting and separation processes. The majority of industrial cutting systems to date having been developed on an iterative experimental basis or by application of a closest match to existing processes. A MAFF LINK funded project into the fundamentals of food fracture and separation showed that food separation situations are complex where interacting 2-D shear stresses can combine to perform cutting more effectively. Initial work developed experimental equipment to measure interrelationships of temperature, cutting type, feed speed and cut parameters in example products such as cheese, beef, and bacon. Cheese was taken to be a homogeneous material but proved not to be so during our trials. The effects of blade friction and effects of temperature on the material properties of cheese had a bigger influence than any other cutting tool or cutting process parameter studied [34].

Further work on robotics and automation in food refrigeration [35] and meat processing [36] continued after moving to Grimsby.

It is interesting to note that some of the first work of the group at Langford was on carcass decontamination. After a hiatus of a decade or so, work on this subject began again in the early 1990s at FRPERC and was expanded to look at many aspects on reducing microbial contamination and improving food safety during primary processing of foods of animal origin and other foods, a number of which looked at the role of temperature. Much of this work (over 30 projects and 70 publications) was funded initially by MAFF and then the UK Food Standards Agency. Many of these projects were carried out in collaboration initially with the Division of Farm Animal Science (DFAS) at the University of Bristol and after moving to Grimsby with the National Centre for Food Manufacturing (NCFM) at the University of Lincoln. Recent and continuing work has been looking at how aspects of the food chain from farm-to-fork may impact on antimicrobial resistance (AMR). Addressing the public health threat posed by AMR is a national strategic priority for the UK. We have recently carried out critical reviews of the evidence on the impact of heat treatments of food on antimicrobial resistance genes and their potential uptake by other bacteria [37] and biocide and metal use during food animal production [38], and further work on AMR is continuing.

Conclusions

Since moving to Grimsby FRPERC produced 107 publications (34 chapters, 18 papers, 40 presentations, 7 articles, and 8 official reports). We worked on 25 publicly funded research projects with 15 centered on food refrigeration. In addition, we continued to carry out many confidential projects for individual refrigeration and food companies, locally, nationally, and internationally; organised training sessions on all aspects of food refrigeration; and carried out legal expert witness activities.

Food refrigeration centred projects	
1	Fostering the development of technologies and practices to reduce the energy inputs into the refrigeration of food (Defra)
2	Review of the risk management practices employed throughout the fish processing chain in relation to controlling histamine formation in at-risk fish species in Scotland (FSAS)
3	Research services/collaboration for investigation into the fundamentals of cryogenic freezing and the evaluation of CAS freezing technology together with its potential when combined with cryogenic freezing
4	Description of the processes used in the UK to manufacture MSM and former DSM meat products from poultry and pork and an initial assessment of microbiological risk (FSA)
5	Qualitative Risk Assessment to support a policy decision on partially eviscerated (effilé) poultry production (FSA)
6	Microbial evaluation of poultry and pork mechanically separated meat (MSM), compared to fresh cuts of meat, meat preparations and minced meat products (FSA)
7	The development of dynamic energy control mechanisms for food retailing refrigeration systems (Innovate UK)
8	Can super-cooling extend the shelf-life of seafood? (Seafish Authority)
9	Feasibility of the IoT for domestic refrigerators, food safety and waste (FSA)
10	Bacteriophage control of listeria in chilled RTE salmon products (Innovate UK)
11	The development of dynamic energy control mechanisms for food retailing refrigeration systems (Innovate UK)
12	The Greater Lincolnshire Agri-Food Innovation Platform (ERDF)
13	The shelf-life of Australian frozen red meat (MLA)
14	Development & optimisation of fresh produce supply chain and storage systems (NEWTON FUND)
15	Quantification of the controls that should be placed on meat prior to mincing (FSA)
Food safety and technology based projects	
1	Creating physical structure from disarray. Developing generic understanding and systems to assess disorder and create physical product arrangements for food industry processing (EPSRC)
2	Study into the use of anal bungs or plugs on pig carcases during processing to reduce faecal leakage
3	Study to optimise the performance and running costs of various commercial scalding systems
4	Study to evaluate decontamination options for automatic and hand scrapers and polishing equipment
5	Study to determine the effectiveness of the application of a second singe on the reduction of surface contamination
7	Exploring the potential for technology to support agency objectives in meat operations (FSA)
8	Decontamination of food: development of the microbial immobilisation treatment of cattle hides (FSA)
9	Assessing the impact of heat treatment on antimicrobial resistance genes and their potential uptake by other 'live' bacteria (FS
10	A critical review of AMR risks arising as a consequence of using biocides and certain metals in food animal production (FSA)

Since April this year the last members of FRPERC have either joined the University of Lincoln or set up an independent consultancy (Fairholme Research Ltd) to continue their activities. We continue to maintain our website FRPERC.com as a repository of much of our past work and publications.

About the authors

Stephen James

Steve joined the Meat Research Institute (MRI) when it was established in 1967 initially involved primarily in research on all aspects of meat refrigeration. He remained at the institute during its change to the Institute of Food Research – Bristol Laboratory (IFR-BL) where his role expanded to cover all foods and heating processes (cooking and decontamination). He established the Food Refrigeration & Process Engineering Research Centre (FRPERC) in 1991 at the University of Bristol when the IFR-BL closed in 1990. He took up a consultancy role at FRPERC when it moved to Grimsby Institute in 2009. He is an author/co-author of over 600 publications on all aspects of the food cold chain from primary cooling through to domestic handling. Although formally retired he still provides support and input to research projects, acts as a reviewer for many food and refrigeration journals and provides expert witness consultancy.

Christian James

Chris worked at FRPERC since graduation in 1993. When FRPERC closed at the University of Bristol in 2009 he moved with some of his colleagues to re-establish FRPERC at the Grimsby Institute. In 2023 FRPERC was closed down at the Grimsby Institute and he established Fairholme Scientific Ltd to continue his work in supporting research and consultancy for the food industry. He is also a Visiting Research Fellow at the University of Lincoln. His research and consultancy have covered many topics such as hygienic control measures during the primary processing of meat and produce, chilling and freezing (including freezing points and supercooling), thawing and tempering, conventional and microwave cooking/pasteurisation, and surface pasteurisation technologies. He has written and presented extensively on these subjects via peer-reviewed papers, conferences, lectures, and books.

References

[1] James, SJ and James, C (2002) *Meat Refrigeration*. Woodhead Publishing Limited.

[2] James, SJ, Purnell, G and James, C (2010) Forty two Years of Food Process Engineering Research at Langford, UK: September 1967 to July 2009. 2nd Edition, available at: FRPERC.com.

[3] Bailey, C and Cox, RP (1976) The Chilling of Beef Carcasses. *Proceedings of the Institute of Refrigeration*, 72, 76-90.

[4] Cox, RP and Bailey, C (1977) Product Loads for Beef Carcass Chilling. *The Proceedings of the Institute of Refrigeration*, 74, 19-28.

[5] Gigiel, AJ and Creed, PG (1987) Effect of Air Speed and Carcass Weight on the Cooling Rates and Weight Losses from Goat Carcasses. *International Journal of Refrigeration*, 10, 305-306

[6] Swain, MVL and James, SJ (1988) Process Design Data for Lamb, Mutton and Goat Chilling. *Meat Chilling*, IFR-BL Subject Day, 23rd February 1988.

[7] Brown, T and James, SJ (1992) Process Design Data for Pork Chilling. *International Journal of Refrigeration*, 15, 281-289.

[8] James, C, Vincent, C, de Andrade Lima, TI and James, SJ (2006) The Primary Chilling of Poultry Carcasses – A Review *International Journal of Refrigeration*, 29, 847-862.

[9] James, SJ, Gigiel, AJ and Hudson, WR (1983) The Ultra Rapid Chilling of Pork. Meat Science, 8, 63-78.

[10] James, C, Briggs, D and James, SJ (2010) The Effect of Freezing and End Cooking Temperature on the Eating Quality of Beef Steaks. *Sustainability & the Cold Chain*, Meeting of IIR Commissions B1, B2, C2, D1 and D2, Cambridge, UK. Paper 211.

[11] James, C, Purnell, G and James, SJ (2015) A review of novel and innovative freezing technologies. *Food and Bioprocess Technology*, 8, 1616-1634.

[12] James, C, Purnell, G and James, SJ (2014) A Critical Review of Dehydrofreezing of Fruits and Vegetables. *Food and Bioprocess Technology*, 7, 1219-1234.

[13] James, C, Reitz, BG and James, SJ (2015) The Freezing Characteristics of Garlic Bulbs (Allium sativum L.) Frozen Conventionally or with the Assistance of an Oscillating Weak Magnetic Field. *Food and Bioprocess Technology*, 8, 702-708.

[14] Rodríguez, AC, James, C and James, SJ (2017) Effects of weak oscillating magnetic fields on the freezing of pork loin. *Food and Bioprocess Technology*, 10, 1615-1621.

[15] Purnell, G, James, C and James, SJ (2017) The effects of applying oscillating magnetic fields during the freezing of apple and potato. *Food and Bioprocess Technology,* 10, 2113-2122.

[16] James, C, Ketteringham, L, Palpacelli, S and James, SJ (2009) Prediction of heat transfer during food chilling, freezing, thawing and distribution. Chpt 4, pp55-78. In *Predictive modelling and risk assessment.*

[17] James, SJ, James, C and Purnell, G (2017) Microwave assisted thawing and tempering. Chpt 12, pp251-272. In *Microwave Processing of Food.*

[18] James, C, Palpacelli, S and James, S (2003) Optimisation of two-stage bacon tempering using mathematical modelling. In *Predictive Modelling in Foods – Conference Proceedings.*

[19] James, SJ (1990) Cooling of cooked products. *Progress in the Science and Technology of Refrigeration in Food Engineering*, Meeting of IIR Commissions, B2, C2, D1, D2/3, Dresden, section 8, 551-557.

[20] Evans, JA, Russell, S and James, SJ (1996) Chilling of Recipe Dish Meals to Meet Cook-Chill Guidelines. *International Journal of Refrigeration*, 19, 79-86.

[21] Ketteringham, L and James, S.J. (2005) Cold Store Design and Maintenance. In 'Handbook of Frozen Food Processing and Packaging', (Ed. Professor Da-Wen Sun), Chapt. 10, pp211-226. ISBN: 157444607X

[22] Cutting, CL, Malton R (1974) Freezing times and weight losses of meat in commercial cold stores. *Proceedings of the Institute of Refrigeration* 70, 75-86.

[23] James, C, Seignemartin, V and James, SJ (2009) The freezing and supercooling of garlic (Allium sativum L.). *International Journal of Refrigeration,* 32, 253-260.

[24] James, C, Hanser, P, Hervier, F and Bouquin, L (2014) Super-cooling Phenomena in Foods. *3rd International ISEKI_Food Conference*, 21-23rd May 2014, Athens, Greece. Book of Abstracts, p73

[25] James, C (2019) Food Transportation and Refrigeration Technologies - Design and Optimisation. Chpt 13, 185-200. In *Sustainable Food Supply Chains: Planning, Design, and Control through Interdisciplinary Methodologies*

[26] Swain, MVL (1997) Retail Display of Unwrapped Meat and Delicatessen Products. In proceedings of *"Meat Refrigeration: Why & How?"*, University of Bristol, Langford, UK, Paper 23.

[27] Russell, SL, Gigiel, AJ and James, SJ (1998) Progress in the Use of Air Cycle Technology in Food Refrigeration and Especially Retail Display. Advances in the Refrigeration System, Food Technologies and Cold Chain, Proceedings of meeting of IIF-IIR Commission B2 & C2, with D1 & D2/3, Sofia, Bulgaria.

[28] Talbot, L, Purnell, G, James, SJ and James, C (2020) Operating Temperatures of Supermarket Frozen Retail Display Cabinets. *International Journal of Refrigeration*, 117, 81-93.

[29] James, SJ and Evans, JA (1992) The Temperature Performance of Domestic Refrigerators. *International Journal of Refrigeration,* 15, 313-319.

[30] James, C, Onarinde, B A and James, SJ (2017) The Use and Performance of Household Refrigerators: A Review. *Comprehensive Reviews in Food Science and Food Safety,* 16, 160–179.

[31] James, SJ, Swain, MJ, Brown, T, Evans, JA, Tassou, SA, Ge, YT, Eames, I, Missenden, J, Maidment, G and Baglee, D (2009) Improving the Energy Efficiency of Food Refrigeration Operations. *Proceedings of the Institute of Refrigeration.*

[32] James, SJ and James, C (2010) The Food Cold-Chain and Climate Change. *Food Research International*, 43, 1944-1956.

[33] James, SJ and James, C (2013) Sustainable Cold Chain. Chpt 20, pp463-496. In *Sustainable Food Processing.*

[34] Brown, T, James, SJ and Purnell, GL (2005) Cutting Forces in Foods: Experimental Measurements. *Journal of Food Engineering*, 70, 165-170.

[35] James, C and James, SJ (2013) Automatic Control of Food Chilling and Freezing. Chpt 12, pp288-303. In Robotics and Automation in the Food Industry: Current and Future Technologies.

[36] Purnell G (2013) Robotics and Automation in Meat Processing. Chpt 13, pp304-328. In *Robotics and Automation in the Food Industry: Current and Future Technologies.*

[38] James, C, Dixon, R, Talbot, L, James, SJ, Williams, N, and Onarinde, BA (2021) Assessing the Impact of Heat Treatment of Food on Antimicrobial Resistance Genes and Their Potential Uptake by Other Bacteria - A Critical Review. *Antibiotics*, 10, 1440.

[39] James, C, James, SJ, Onarinde, BA, Dixon, RA, and Williams, N (2023) A Critical Review of AMR Risks Arising as a Consequence of Using Biocides and Certain Metals in Food Animal Production. *Antibiotics*, 12, 1569.

Q&A- The History of the Food Refrigeration and Process Engineering Research Centre (FRPERC) and its Role in Food Refrigeration Research

Dermot Cotter – I know in the Cold Chain Federation there is a campaign for a 3 degree increase from -18 to -15. Do you believe it is just for meat that the temperature can increase or is it across all sectors?

Steven James – I'm certain there are many product sectors that you do that. We've done a lot of work looking at fish, it's mainly legal work looking at problems with storage of fish. We have had fish that is stored well above the temperature that people think, up to -12 and after 2 or 3 years the quality of that fish has been better than some of the stuff that's stored at lower temperatures. So certainly you could raise the temperature with fish. Chris, probably fruit and vegetables as well?

Christian James – The interesting thing on temperature is how long you are going to store it and what the temperature life of things are. The literature is interesting. There are arguments about the length of storage and how that affects the temperature, short term storage can be done at a warmer temperature and long term storage needs a colder temperature. -18 came in particularly with the development of frozen vegetables and fruit and maintaining the nutritional quality for over a year with no change in vitamin or eating quality. That may not necessarily be what everybody is after. From a hygiene and safety view pathogens can't grow much below -1 and even yeasts and moulds can't grow much below -8.

DC – What would be the difference in refrigeration effect by putting a door on a retail cabinet?

SJ – It would have a fantastic effect. Most retailers accept the fact that it is beneficial to put doors on. You're going to save at least 10 – 20% depending on where you are. If doors are opened a lot then there will be less saving.

DC – Where are the gaps left in this industry to research?

SJ – I think there are gaps everywhere. The worry is that I'm not sure who is filling these gaps.

CJ – For a long time there have been a lot of fixed temperatures of where you go to. Maybe with dynamic shelf life there's a realisation that there's a whole range of temperatures that can be applied.

SJ – It worries me about AI applying to food refrigeration because it is going to say fast freezing good, slow freezing bad because the literature says that in the past but it's not actually true.

The knowledge hub for refrigeration, air conditioning and heat pumps

Waste heat from the London Underground: an investigation of the potential benefits of integrating heating and cooling

Henrique Lagoeiro, *Gareth Davies, Akos Revesz, Daniel Curry, Josh Vivian, Gareth Faulks, Ken Gysin, Declan Murphy and Graeme Maidment FInstR*

Henrique Lagoeiro
Winner of the 2022/23 Ted Perry Award

Why you should attend

- Learn how modelling was used to develop a heat recovery network that provide s a low-carbon solution to heating buildings and cooling underground trains
- Understand the carbon and cost benefits of implementing a networked heating and cooling system
- See the results of the project measured in terms of both heating and cooling achieved

Abstract

Waste or recoverable heat is a resource that could be cost-effectively exploited by district heating (DH) systems to increase the efficiency of heat supply in buildings. This paper introduces the Bunhill waste heat recovery (WHR) system, a novel scheme that recovers waste energy from a ventilation shaft of the London Underground (LU) transport network. The system is based upon the installation of a heat recovery heat exchanger consisting of cooling coils and a reversible fan; the coils are connected to a heat pump that supplies low-carbon thermal energy to the Bunhill DH network in central London. One particularly important aspect of the Bunhill WHR system is its ability to operate in a way that not only provides heating to the local heat network, but can also simultaneously supply cooled air to the LU tunnels depending on the operation of the reversible fan. The current paper presents the results from an investigation into the benefits of integrating heating and cooling from the perspective of reducing carbon emissions and the levelised costs of energy from the WHR system, whilst also alleviating peak temperatures at nearby LU stations through the cooling provided. The findings of the investigation are presented together with recommendations for further development and future deployment of WHR systems, which not only apply to underground railways but also to other sources of recoverable heat.

Keywords

Heating and cooling, heat pumps, waste heat, heat networks, district energy, London Underground, railway tunnels, modelling

1.Introduction

1.1 Background

Climate change represents an existential threat to mankind that has prompted many countries worldwide to commit to achieving carbon neutrality by the middle of this century. In many cases, net-zero targets have been influenced by the remarkable progress made by the electricity sector; for instance, over the last decade, the UK has experienced a rapid decarbonisation of the electricity grid, with renewables accounting for a record high of 43.1% of the electricity generated in 2020 [1]. This led to a fall in national GHG emissions to 48.7% of their 1990 level [2]. Despite this great progress, recent crises have emphasised some of the vulnerabilities with our energy system, which must be addressed as we start to tackle hard-to-abate areas, such as the built environment.

For example, the post-pandemic economic recovery and the war in Ukraine have triggered a cost-of-living crisis in the UK due to a surge in natural gas prices, with inflation reaching its highest rate in 40 years [3]. This also affected the electricity market, as the principle of marginal cost pricing makes fossil fuels predominant in setting the wholesale price of electricity [4]. It has been estimated that the 2021/22 energy crisis has increased the number of fuel-poor households in the UK by around 68% [5]. This scenario has highlighted the key role of energy efficiency, which can help to drive down demand and reduce our reliance on fossil fuels. Even as we move towards electrification, energy efficiency will remain instrumental, particularly in terms of reducing grid upgrading and reinforcement costs.

Modelling from DESNZ projects that peak electricity demand will increase from 58 GW in 2020 to 190 GW in 2050. In a net-zero scenario, the necessary grid investment to meet that demand in 2050 varies from £270-350bn, meaning that each GW of additional demand increases network investment costs by £2-2.65bn [6].

A recent review has gathered evidence on the unsuitability of hydrogen as an alternative for natural gas for heating homes, with electrification being highlighted as an already available solution for decarbonising heat in buildings [7]. The electrification of heat remains, however, a significant challenge; a 2018 report showed how the peak gas demand in that year was 214 GW, nearly 4 times higher than the electricity demand of 53 GW at the same time [8]. This also reinforces the relevance of energy efficiency in the transition to low-carbon heating. When it comes to buildings, energy efficiency is often discussed from the perspective of improving the building stock, but efficient heat supply is equally important. One opportunity for enhancing system-wide efficiency is to capture and reuse waste (recoverable) heat, which is normally dissipated to the environment. This wasted thermal energy can be recovered and delivered to end users through district heating (DH), effectively reducing primary energy demand, thus tackling issues around both carbon emissions and energy security.

1.2 Waste heat from the London Underground

DH networks have been increasingly identified as a key technology for the decarbonisation of buildings at scale. DH is best suited for densely populated urban areas, being particularly competitive when waste or renewable heat is available nearby. The opportunity for integrating waste heat and large-scale heat pumps has been predominant in recent generations of DH. The earliest DH generations used high-temperature (i.e. 100°C and above) steam and water, e.g. 1st and 2nd generation (1G and 2G) DH networks [9]. However, operating temperatures have steadily decreased for successive generations of DH i.e. 3G, 4G and 5G networks, with the latest (5G) networks operating at close to ambient temperature [10]. Typically, 4G networks operate with supply temperatures of 50-60°C, whilst 5G systems operate at a lower supply temperature range (15-25°C). The lower distribution temperatures associated with the 4th and 5th generations increase the economic potential for using waste heat, which can be either used directly or after upgrade by a heat pump.

For this reason, opportunities for WHR at low temperatures (<100°C) have been increasingly reported in the literature in recent years, particularly with the advent of 4G and 5G networks. There are many sites in urban settings from which it is possible to capture waste heat, such as industrial plants, data centres, electricity distribution systems, sewers and supermarkets. As discussed in [11], another source of great potential is underground railways, particularly for large cities like London, where public transportation relies heavily on a metro system. The LU network has a total length of 402 km, of which 45% is in tunnels, and carries around 1.35 billion passengers annually [12]. In 2016/2017, the Underground consumed over 1,700 GWh of electricity, with around 500 GWh of energy ending up degraded and released as waste heat [13]. These conditions lead to tunnel temperatures typically varying between 18 and 28°C throughout the year. The current thermal environment of the LU means that there is significant potential for recovering waste heat from the tunnels, while cooling solutions are expected to gain importance as air temperatures rise in the future due to climate change.

1.3 Critical infrastructure and cooling in a warming world

The cooling requirements of critical urban infrastructure is expected to grow massively in the coming decades. UNEP estimates that a business-as-usual scenario with current policies is likely to lead to an increase of 2.8°C in average surface temperatures by the end of the century [14]. The UK is already experiencing extreme heat waves, with temperatures exceeding 40°C [15], which pose significant challenges to the operation of services such as mass transport systems. The Birmingham Energy Institute argues that there is a compelling case for a formal designation of cooling systems as critical infrastructure [16]. A recent investigation has highlighted climate-related risks to the operation of LU's services, which include issues like asset failure and exposure of passengers and staff to heat-related illnesses. This is likely to increase maintenance requirements and train service delays in the future [17]. With a growing need for sustainable heating and cooling in cities, solutions like the Bunhill WHR system represent an efficient way of tackling both challenges with the same infrastructure

2. The Bunhill waste heat recovery (WHR) system

The WHR system was introduced into the existing Bunhill DH network as part of an extension project known as Bunhill 2. The system recovers waste heat from the City Road ventilation shaft, located on the Northern Line between Angel and Old Street stations in central London. The main components of the WHR system are a reversible fan, heat recovery coils with a nominal capacity of 780 kW, a 1 MW two-stage ammonia heat pump, and a coolant loop that connects the coils to the heat pump. The WHR system was constructed within a new energy centre, which also houses a 50 m^3 thermal energy store and two 237kW$_e$/372kW$_{th}$ combined heat and power (CHP) units. These additional heat sources, together with the energy centre from the initial network, enable the system to operate flexibly while meeting a heat demand associated with 1,350 dwellings, two leisure centres and a local primary school. The heat network operates with flow and return temperatures of 75 and 55°C, respectively.

An essential feature of the Bunhill WHR system is its ability to supply cooled air to the LU tunnels whilst simultaneously delivering heat to the local heat network, depending on the direction in which the reversible fan operates. If operating in extract mode, the system utilises tunnel air as the heat source and no cooling is provided to the LU network. However, when operating in supply mode, the system recovers heat from ambient air, which is cooled down in the process before being supplied to the tunnels. Both extract and supply modes are illustrated in Figure 1.

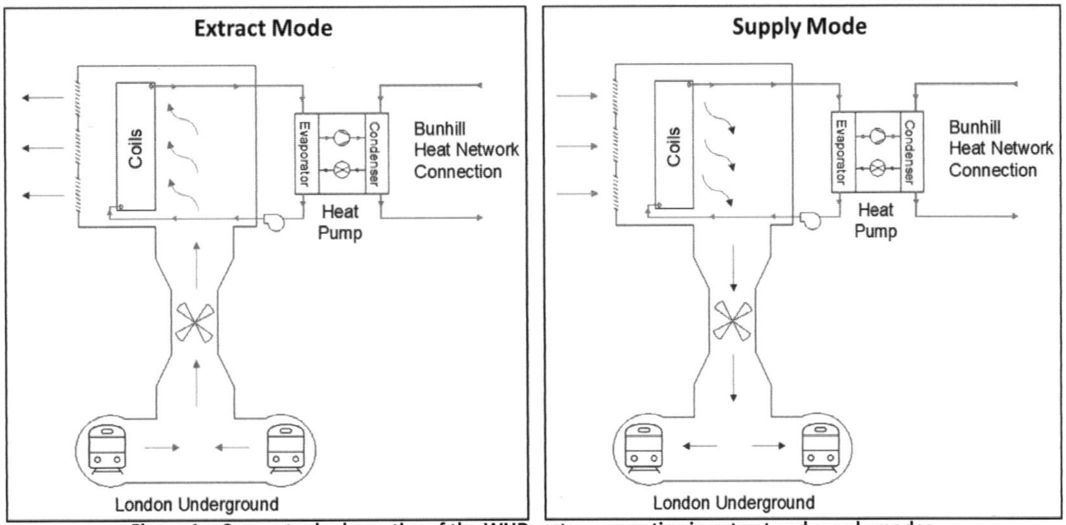

Figure 1 – Conceptual schematics of the WHR system operating in extract and supply modes.

As the main novelty of Bunhill 2 is associated with WHR from the LU, this paper focuses on investigating how a large-scale heat pump utilising waste heat would perform against conventional technologies, namely communal gas boilers and a low-carbon alternative of individual air-source heat pumps (ASHPs). In order to realise the full benefits behind the WHR system, the impacts of cooling will also be evaluated. This includes calculating overall cost and carbon benefits when operating in supply mode, as well as investigating platform temperature reductions that could be achieved at nearby stations in 2030.

3. Modelling of the WHR system

The performance of the WHR system was simulated using a mathematical model developed with the commercial software tool Engineering Equation Solver (EES). The WHR model was introduced in [18] and is capable of iteratively solving thermodynamic balance equations across the system, being used to determine its energy consumption, as well as heating and cooling outputs. The energy consumption calculations are associated with the electricity used to run the reversible fan (W_F), the coolant circulation pump (W_p) and the heat pump (W_{HP}) when delivering a heat output of Q_{out}. Based upon design specifications, as well as inputs for air temperature and humidity, the model is able to calculate the heat recovered by the heat exchanger (Q_{rec}), which is assumed to be equal to that absorbed by the heat pump's evaporator ($Q_{in} = Q_{rec}$). When operating in supply mode, the heat recovered is also equal to the cooling that is delivered by the WHR system. A schematic of the main energy inputs and outputs of the model is shown in Figure 2.

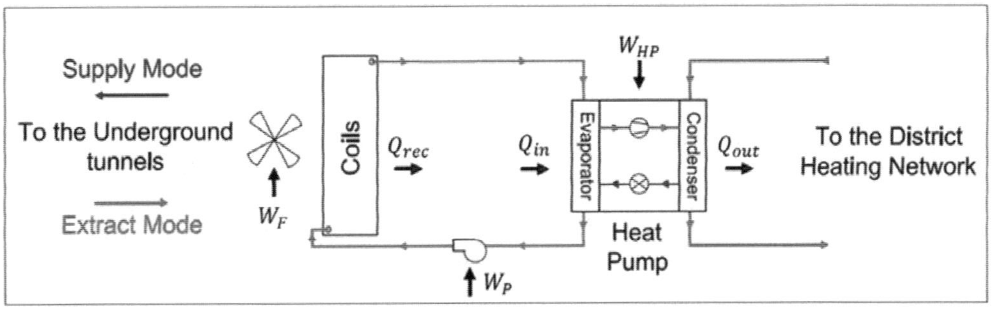

Figure 2 – Schematic highlighting energy inputs and outputs associated with the WHR model.

Each of the components illustrated in Figure 2 together with its outputs were modelled by solving the mass and energy balance between the coils and the heat pump. These outputs are illustrated in Figure 3, along with the main inputs and connections between the different model components. The calculations were carried out on an hourly basis, using temperature and relative humidity recordings from the ventilation shaft and the nearest available weather station, which were provided by Transport for London (TfL) and the UK's Meteorological Office, respectively, for the period from January 2013 to January 2014. The two-stage heat pump was modelled according to the design specifications for its compressors and heat exchangers, and the coils were simulated considering the latent and sensible heat transfers that take place when heat is recovered from humid air. The WHR model also calculates the energy consumption for the pump and fan based on the expected pressure drops that they must overcome. The model was validated against manufacturer's data and a detailed description of each modelled component can be found in [18]. By modelling each component, the coefficient of system performance (*COSP*) can be calculated, as shown in Equation 1.

$$COSP = \frac{Q_{out}}{W_{HP}+W_F+W_P} \qquad [1]$$

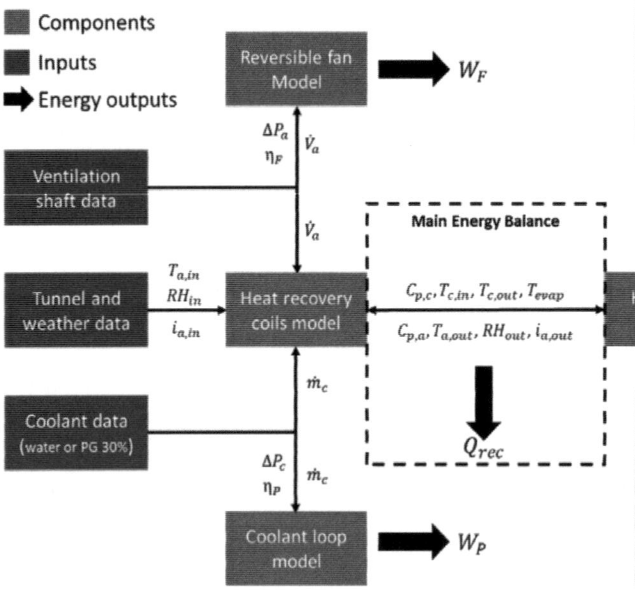

Nomenclature	
C – Specific heat (KJ/kg-K)	RH – Relative humidity (%)
i – Enthalpy (kJ/kg)	T – Temperature (°C)
$\dot{m}$ – Mass flow rate (kg/s)	$\dot{V}$ – Volumetric flow rate (m³/s)
P – Pressure (kPa)	W – Work (kW)
Q – Heat (kW)	η – Efficiency (%)

Subscripts	
a – Air	HP – Heat pump
c – Coolant	P – Pump
$evap$ – Evaporating	p – At constant pressure
F – Fan	PG – Propylene glycol
in – Inlet	r – Recovered
HP – Heat pump	w – Water

Figure 3 – Framework for the WHR model, highlighting components, inputs and outputs.

4. The case study

The Bunhill DH network has been used as a case study for evaluating the benefits of recovering and reusing waste heat from the London Underground. This involved an analysis of how the WHR system could meet the heat demands of the buildings connected to the DH network, which are illustrated in Figure 4. For this purpose, hourly heat demand profiles for the residential buildings connected to the network were provided by Islington Council and used to assess the levelised costs of meeting part of the network heat demand with a 1 MW heat pump utilising waste heat from the London Underground

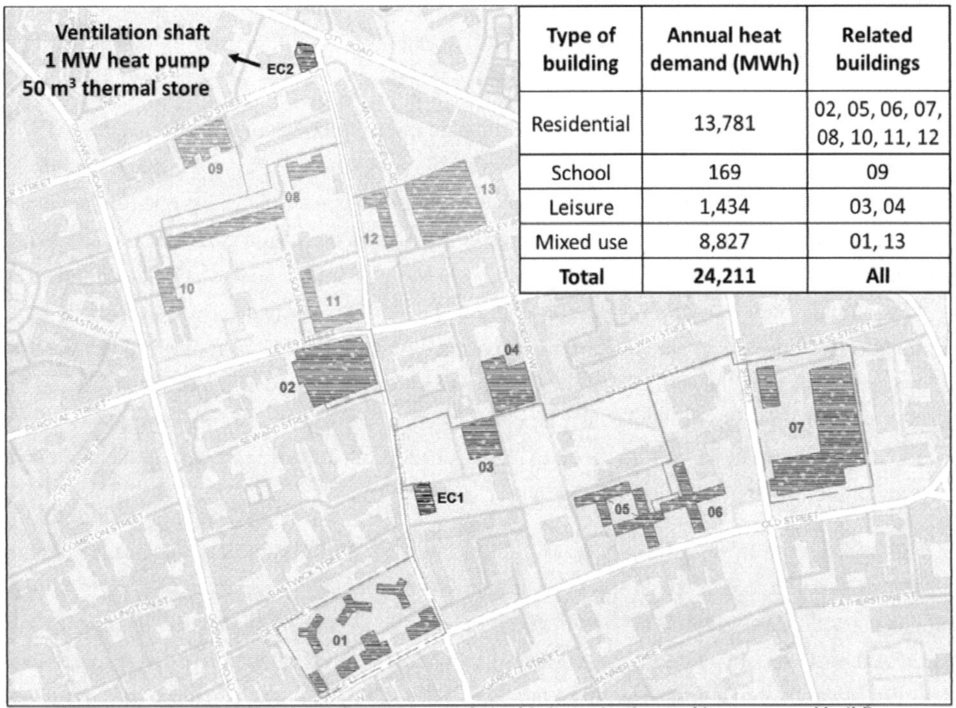

Type of building	Annual heat demand (MWh)	Related buildings
Residential	13,781	02, 05, 06, 07, 08, 10, 11, 12
School	169	09
Leisure	1,434	03, 04
Mixed use	8,827	01, 13
Total	**24,211**	**All**

Figure 4 – The DH network used as a case study in this investigation and its connected buildings.

The council estates connected to the network have a combined heat load of 13,781 MWh per annum. The hourly heat demand profiles were matched against the potential heat output from the WHR system, which enables analysing how much of the network demand could be met with the 1 MW heat pump. As the two-stage heat pump has variable speed compressors, it was assumed that it was able to vary its thermal output from 30% to 100% of its capacity (300 kW to 1 MW) without affecting the COSP. The 50 m³ thermal store was also modelled to provide additional flexibility to the WHR system's operation. A heat loss rate of 0.6% per hour was assumed for the thermal storage tank [19]. The DH network was assumed to have heat distribution losses of 10%, which is in accordance with the maximum expected loss rate for new systems [20].

As for the cooling benefit, it was modelled by considering that cooling would be delivered when the system operates in supply mode, and five scenarios involving different combinations of extract and supply modes were modelled, as shown in Table 1. For scenarios when the fan was running exclusively in extract mode, the WHR system was modelled with water as the coolant. However, when supply mode operation was also required, an anti-freeze mixture needed to be utilised and a 30% mixture of propylene glycol (PG) and water was considered. Propylene glycol was chosen due to its extremely low environmental, health, fire and corrosion risks [21]

Table 1 – Different modelling scenarios for the WHR model, based upon fan operation mode.

Scenario	Operation mode	Description
1	12E/0S	Fan operating in extract mode for the entire year (12 months).
2	9E/3S	Fan operating in supply mode during meteorological summer (Jun/Jul/Aug), and in extract for the rest of the year.
3	6E/6S	Fan operating in supply mode for half the year, from May to October, and in extract for the remaining 6 months.
4	3E/9S	Fan operating in extract mode only during meteorological winter (Dec/Jan/Feb), and in supply for the rest of the year.
5	0E/12S	Fan operating in supply mode for the entire year

5. Modelled operation of the DH network

The model was first applied to evaluate how much of the heat generated by the WHR system could be effectively used by the local DH network, and the results of this analysis are shown in Table 2. The modelled scenarios reflect the operation modes described in Table 1. Due to data availability, only the residential heat demand of the Bunhill DH network was considered in the modelled scenarios. Based on the hourly heat demand profiles, the WHR system would produce 7,773 MWh of heat annually, meeting a demand of 6,996 MWh per annum after losses, which is equivalent to 50.8% of the total heat demand considered. In this case, the thermal store would play a minor role, storing only 1% of the total heat produced by the WHR system. Throughout the year, the WHR system would be meeting the base load of the DH network, and the thermal store would mainly be used by the heat pump in summer months, when demand on the network is low. However, storage capacity can still be useful to manage heat production from CHP or if the heat pump is operated with time-of-use tariffs, as investigated in [22].

As it can be observed from Table 2, the outputs and efficiency of the WHR system vary significantly depending on the mode of operation considered. In scenarios predominantly running on extract mode, more waste heat is used and higher heating efficiencies are obtained. While scenario 1 obtained the highest heating *COSP* of 3.36 and a waste heat utilisation rate of 89%, scenario 5 would only use ambient air as its heat source (waste heat utilisation rate of 0%), therefore achieving a heating *COSP* of 2.65. However, when cooling outputs are also considered, the efficiency of the WHR system is higher for scenarios where supply mode is used for most of the year. A combined heating and cooling *COSP* of 4.30 was obtained for scenario 5, when the cooling output would be 62% of the total heating output from the WHR system. On the other hand, the combined heating and cooling efficiency of scenario 1 would be the same as its heating-only efficiency, as no cooling is provided when the WHR system operates only in extract mode.

Table 2 – Modelled operational parameters for the WHR system in the different scenarios described in Table 1.

WHR system and DH network parameters	Scenarios				
	1	2	3	4	5
Heat generated by WHR system annually (MWh)	7,773	7,773	7,773	7,773	7,773
Heat delivered annually, excl. distribution losses (MWh)	6,996	6,996	6,996	6,996	6,996
LU waste heat recovered and reused annually (MWh)	5,457	4,468	2,966	1,483	0
Cooling delivered annually by WHR system (MWh)	0	896	2,314	3,583	4,842
Annual WHR system energy consumption (MWh)	2,316	2,410	2,494	2,708	2,931
Seasonal COSP for WHR system (heating only)	3.36	3.23	3.12	2.87	2.65
Seasonal COSP for WHR system (heating and cooling)	3.36	3.60	4.04	4.19	4.30
Ratio of cooling to heating production	0.00	0.12	0.30	0.46	0.62
Share of available waste heat reused annually (%)	89%	74%	50%	26%	0%
Share of annual heat demand met with waste heat (%)	40%	32%	22%	11%	0%

6. Integrated benefit analysis

The integrated heating and cooling benefits of the WHR system were investigated by applying two key metrics. The first one is the levelised cost of energy (LCOE), which is commonly used to express the full costs of energy supply in unitary terms, and is applied in this investigation by considering both heating and cooling supplies. The second metric is the carbon abatement cost (CAC), which represents the cost-effectiveness of a mitigation measure and can be applied to compare different technologies in terms of the cost per metric tonne of CO_2 equivalent (tCO_2e) avoided over a given period [23].

6.1 Levelised cost of energy (LCOE)

The LCOE can be calculated as shown in Equation 2, where Q_a and C_a represent, respectively, the annual heating and cooling delivered by the WHR system, excluding losses. $CAPEX_a$ and $OPEX_a$ are the annualised capital and operational expenditure associated with the system. The $CAPEX_a$ was calculated separately for each system component (i.e. WHR system, TES and DH connection), considering infrastructure design life (L), as well as an assumed loan repayment period (t) and annual interest rate (i_a), as shown in Equation 3. As for $OPEX_a$, it combines annual variable ($O\&M_v$) and fixed ($O\&M_f$) operational and maintenance costs, as shown in Equation 4.

$$LCOE = \frac{CAPEX_a + OPEX_a}{Q_a + C_a} \quad [2]$$

$$CAPEX_a = \frac{CAPEX\,(1+i_a)^t}{L} \quad [3]$$

$$OPEX_a = O\&M_v + O\&M_f \quad [4]$$

$$CAPEX_{TES} = 8357 \times V_{TES}^{-0.47} \quad [5]$$

In all scenarios, the loan period for the investment costs was assumed to be 10 years with an interest rate of 3.5% [24], and all the assumptions used in LCOE calculations are listed in Table 3. In order to reflect economies of scale, the CAPEX benchmark for the thermal store ($CAPEX_{TES}$), in GBP/m^3, was correlated to its volume (V_{TES}), in m^3, using Equation 5, obtained from the Danish Energy Agency [25] and corrected to account for inflation. The cost benchmarks were mainly obtained from the Danish Energy Agency's technology catalogue [25], and values shown in Table 3 are the final benchmarks in GBP after conversion from EUR, considering the inflation and exchange rates also shown in Table 3. LCOE values were then calculated for each of the scenarios listed in Table 1. The results obtained were then compared to the operational costs of meeting the same heat demand with either communal gas boilers or individual ASHPs. The calculations consider both O&M and energy costs for communal boilers and only the energy costs for running ASHPs. These were calculated by considering an 85% efficiency for gas boilers and a seasonal COP of 2.68 for ASHPs [26], and the latest energy tariffs published by DESNZ [27]. In the case of communal gas boilers, non-domestic prices would apply as the heating plant would be operated by the local council, but individual ASHPs were assumed to run on domestic electricity prices, whereas the WHR system would be able to take advantage of lower tariffs for non-domestic electricity.

Table 3 – List of assumptions used in levelised cost of energy calculations and their references.

Assumption	Value	Unit	Reference
Capital costs			
Assumption	**Value**	**Unit**	**Reference**
Heat recovery equipment for ventilation shaft	975,780	GBP/MW	[28]
District-scale heat pump using waste heat	1,390,957	GBP/MW	[25]
Heat exchanger substation	115,043	GBP/MW	
Pumping station	103,538	GBP/MW	
Pipework for DH connection	1,170	GBP/m	[29]
Operational costs			
Assumption	**Value**	**Unit**	**Reference**
Annual district heating O&M fixed costs	N/A	GBP/MWh	[25]
Annual district heating O&M variable costs	1.62	GBP/MWh	
Annual thermal storage tank O&M fixed costs	56	GBP/unit	
Annual thermal storage tank O&M variable costs	0.79	GBP/MWh	
Annual gas boiler O&M fixed costs	2,110	GBP/MW	
Annual gas boiler O&M variable costs	1.19	GBP/MWh	
Non-domestic price of natural gas	65.50	GBP/MWh	[27]
Non-domestic price for electricity	248.1	GBP/MWh	
Domestic price for electricity	336.2	GBP/MWh	
Financial figures			
Assumption	**Value**	**Unit**	**Reference**
Inflation coefficient from 2015 to 2023	129	%	[3]
Inflation coefficient from 2020 to 2023	117	%	
Loan repayment period	10	years	N/A
Loan interest rate	3.5	% p.a.	[24]
Technology design life			
Assumption	**Value**	**Unit**	**Reference**
Waste heat recovery system	25	years	[25]
District-scale heat pump	25	years	
Small-scale thermal storage tank (steel)	30	years	
District heating network	50	years	

A comparison of the levelised costs for the WHR system against the operational costs for gas boilers and ASHPs is shown in Figure 5. The contributions of different costs to the final levelised figure of each scenario are also indicated. As it can be observed, an increase in the amount of cooling delivered would achieve significant reductions in the LCOE, as the useful energy outputs from the system increase considerably when both heating and cooling are exploited, with scenario 5 obtaining the lowest LCOE value of £76.7 per MWh, which is 2% and 39% lower than the operational costs of communal gas boilers and ASHPs, respectively. Scenario 1, which corresponds to year-round operation in extract mode, uses less energy due to its higher seasonal COSP, as shown in Table 2, but its lack of cooling provision led to the highest LCOE (37% higher than the OPEX for gas boilers). However, the LCOE was still 14% lower than the running costs of an ASHP.

Despite these results indicating a benefit of running the system in supply mode throughout the year, one potential issue is that the recipient of the cooling benefit would be the railway operator, whilst the DH operator would be the stakeholder having to bear the higher costs of producing low-carbon heat using electricity as fuel. This might pose a challenge to long-term cooling delivery, but the results from Table 2 still indicate how it is possible to deliver cooling when it is most needed with a small impact on the annual energy consumption of the WHR system. Furthermore, the high CAPEX associated with DH networks still leads to long payback periods, meaning public policy must play a role as an enabler for WHR projects, especially as there is a significant disparity between gas and electricity prices in the UK. The higher levies and taxes applied to electricity bills represent a risk for the electrification of heat supply and for the exploitation of low-grade waste heat sources in the UK. Government plans to rebalance gas and electricity costs can be key in accelerating the rollout of DH networks using waste heat, together with policies such as the Green Heat Network Fund [22], which can cover up to 50% of the total commercialisation and construction costs for low-carbon DH projects.

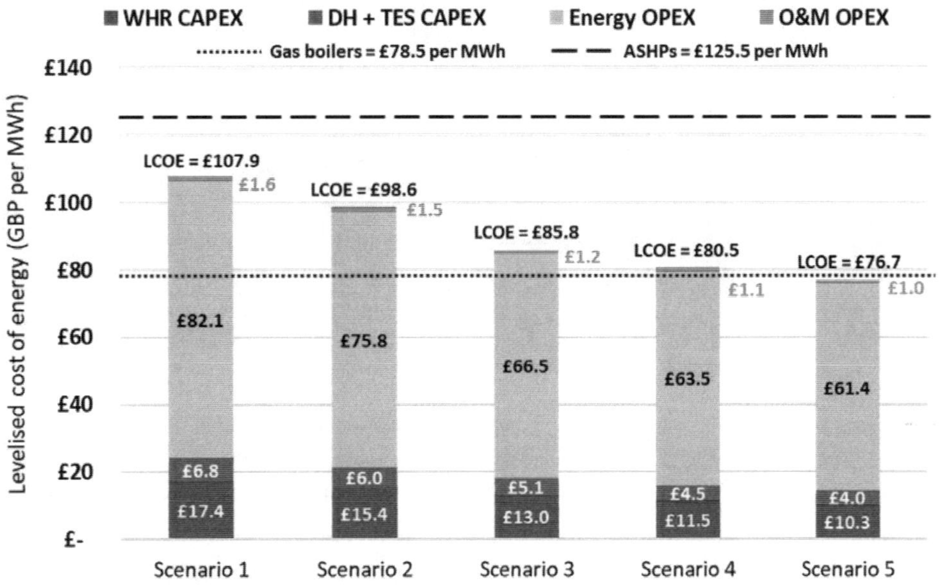

Figure 5 – Levelised cost of energy comparison between each modelled scenario and conventional technologies

6.2. Emissions savings and carbon abatement cost (CAC)

The environmental benefits of reusing recovered waste heat were assessed using the CAC metric, which can be interpreted as an indication of the decarbonisation costs associated with a given low-carbon technology. The CAC values were calculated for both the WHR system and individual ASHPs using communal gas boilers as the business-as-usual reference case. The CAC can be calculated as shown in Equation 6, which includes the net annual cost, i.e. the difference between the annualised costs for the low-carbon alternative ($C_{low,carb}$) and the operating costs of the displaced incumbent technology ($C_{inch,tech}$), as well as the carbon emission savings (S_{CO2}) achieved after the low-carbon technology is introduced. These were calculated considering carbon intensity factors for electricity and natural gas of 0.129 and 0.183 tCO_2e per MWh, respectively [29].

$$CAC = \frac{c_{low,carb} - c_{inc,tech}}{s_{CO_2}} \quad [6]$$

The carbon emissions savings and CAC values for each modelled scenario are illustrated in Figure 6. The carbon savings for WHR include both heating and cooling and are shown in Figure 6(a). Cooling emission savings were estimated by considering that the WHR system would displace a chiller with a COP of 2.70, which is based on an existing system installed at a ventilation shaft on the Victoria Line. In general, carbon emission savings increase for higher shares of supply mode operation, when greater amounts of cooling are delivered, increasing by 242% for scenario 5 (2,920 tCO_2e) when compared to scenario 1 (1,206 tCO_2e). However, it is unlikely that vent shaft chiller systems would operate for the entire year, so the savings in scenario 5 must be interpreted as an upper limit rather than the actual savings that would be achieved in practice. If only heating is considered, carbon savings would be 7% lower for scenario 5 in comparison to scenario 1, being 4% lower than the savings achieved with ASHPs. For a year-round operation in extract mode, the WHR system would require 13% less energy than individual ASHPs due to its higher COPs, equally impacting operational carbon emissions.

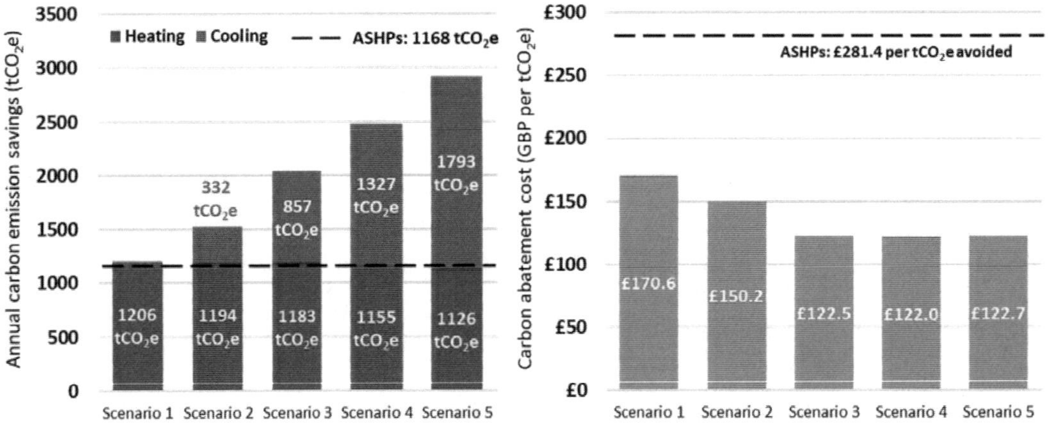

Figure 6 – Annual carbon emission savings (a) and CAC values (b) for all modelled scenarios, compared against ASHPs.

The results from the CAC analysis are illustrated in Figure 7(b). The WHR system was able to achieve much lower CAC values when compared to ASHPs (from 43% to 61% of the ASHP value), which is due to the higher efficiencies and greater carbon savings of WHR across all scenarios. It is interesting to note that, despite the highest emission savings for scenario 5, its high annual costs (20% greater than scenario 1, see Table 2) meant it would not achieve the best results in terms of CAC, which was obtained for scenario 4 (£122 per tCO_2e). CAC values were only marginally different for scenarios 3, 4 and 5, indicating how scenario 3, which involves operating for only half the year in supply mode, would achieve a low decarbonisation cost without overburdening the DH network operator with a lower system efficiency. The results from this analysis highlight how WHR systems can lead to significant benefits in terms of reducing the costs of decarbonisation, particularly when additional value streams such as cooling are also considered.

7. Cooling the London Underground

The cooling benefits behind the WHR system are also assessed in terms of the potential reduction in platform temperatures that can be achieved at the nearest stations to the ventilation shaft. This investigation was carried out in collaboration with the engineering team at TfL in order to simulate how future network temperatures would be affected by the WHR system, utilising a bespoke modelling tool based upon the Subway Environment Simulation (SES) platform.

SES is able to perform 1D simulations of the operation of trains in tunnels, being suited to model many different aspects of a subway environment, such as airflows, temperatures and humidity throughout stations, tunnels and ventilation shafts [32]. The thermodynamic simulations in SES consist of breaking down the network into smaller components of constant temperature and humidity. The heat generated within each component over time, based upon train profiles and airflow patterns, is then used to calculate energy and mass balances at nodes connecting subsequent components, whilst also taking into account the conductive heat transfer between tunnel walls and the surrounding soil [33]. This approach is used for both aerodynamic and thermodynamic calculations, and an example of how different network sections are modelled is shown in Figure 7.

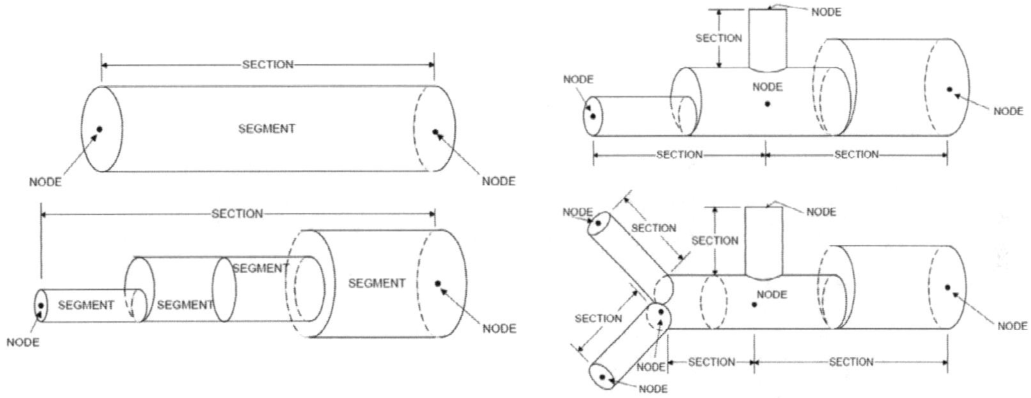

Figure 7 – Examples of network sections and their components as simulated in SES [33].

The results from the WHR model were used as inputs for SES, enabling the analysis of how the stations of Angel, Old Street, King's Cross and Moorgate would be influenced by the provision of cooling for the Table 1 scenarios over the long term, considering 2030 as the target year. The simulations work by adjusting the dry and wet-bulb temperatures of the air that is supplied at the ventilation shaft node to the conditions predicted by the WHR model. The novelty behind this approach is enabling the use of an accurate representation of the cooling process, which is achieved with the WHR model, to investigate the impacts of cooling coils on the LU environment.

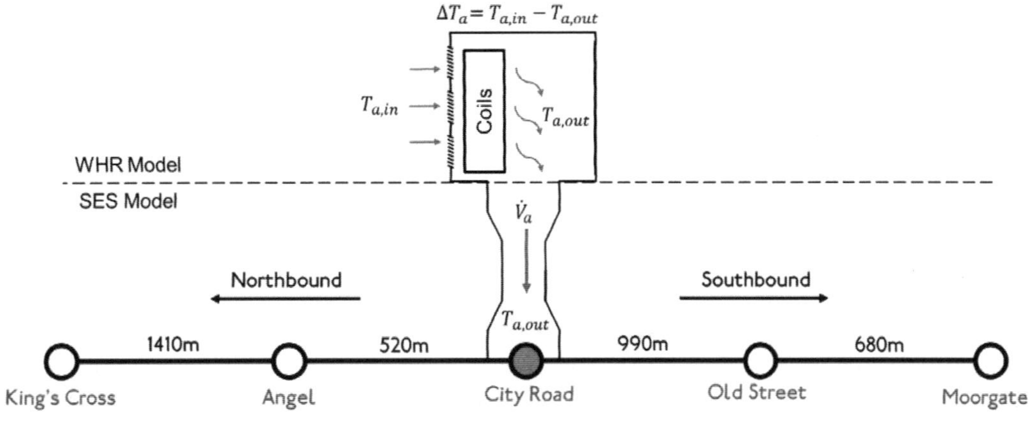

Figure 8 – Schematic highlighting the inputs from the WHR model to the SES simulations.

The SES model developed by TfL is calibrated to utilise 2006 weather data as the basis for simulations, and the UK climate projections (UKCP) from 2009, along with train frequency profiles, are utilised to yield future platform temperatures. Therefore, the WHR model had to be run with 2006 weather data in order to provide the necessary inputs for this investigation, and the link between the WHR and SES models are provided in Figure 8. The cooling effect calculated by the WHR model considers both sensible and latent cooling loads associated with the heat recovery process in supply mode, as discussed in [18], and the data used for simulation are summarised in Figure 9.

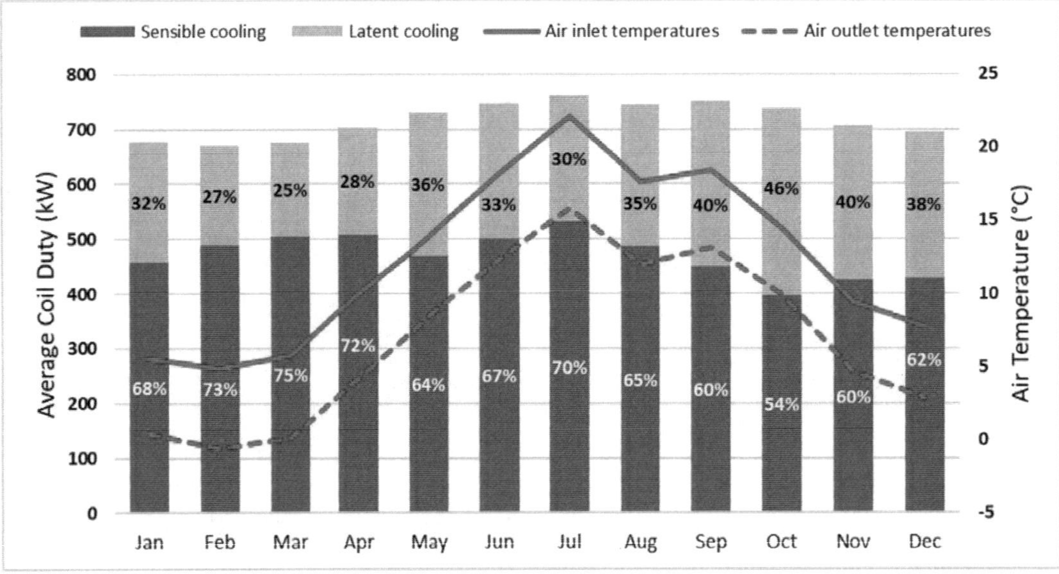

Figure 9 – Monthly average sensible and latent cooling outputs, as well as air inlet and outlet temperatures calculated by the WHR model in supply mode.

As shown in Figure 9, supply mode operation results in significant proportions of latent cooling, with an average of 34% of the total coil duty annually. This leads to a lower air temperature reduction (ΔT_a) than would be achieved if no condensation took place. Overall, the annual average ΔT_a was calculated as 5.4°C, although values as high as 9.6°C were predicted for an entirely sensible cooling process, highlighting the relevance of considering latent loads when analysing the cooling potential of the WHR system. For each scenario from Table 1, supply mode operation was modelled using the air outlet temperatures corresponding to the months the system would be operating in supply mode. Extract mode was modelled with no air supply through the vent shaft, and a mixed operation was simulated by combining the results from both extract and supply simulations. The results are expressed in terms of average evening peak temperatures for the hottest week of the year. This corresponds to the worst-case scenario the stations would be exposed to, representing how the cooling provided can alleviate temperatures during critical periods. The results from the SES analysis are illustrated in Figure 10.

The results show how the scenarios involving cooling provision would lead to significant reductions in platform temperatures as opposed to year-round extract mode operation, particularly for the stations adjacent to the ventilation shaft (Angel and Old Street), as negligible reductions were observed at King's Cross and Moorgate. The highest reductions were estimated for scenario 5, where the year-round supply of cooling could potentially reduce peak temperatures by 7.2 K at Angel and 6.3 K at Old Street. .For scenarios 2, 3 and 4, which involve a combination of extract and supply modes, the average ΔTs, considering both adjacent stations, were of 1.1, 2.6

and 4.5 K, respectively, highlighting how the cooling benefit can be increased if the system operates for longer periods in supply mode. These temperature reductions might lead to several tangible benefits for LU, such as increasing the wellbeing of passengers and staff [34], reducing risk of train delays caused by high temperatures [17], as well as unlocking potential for service frequency and ridership to be increased.

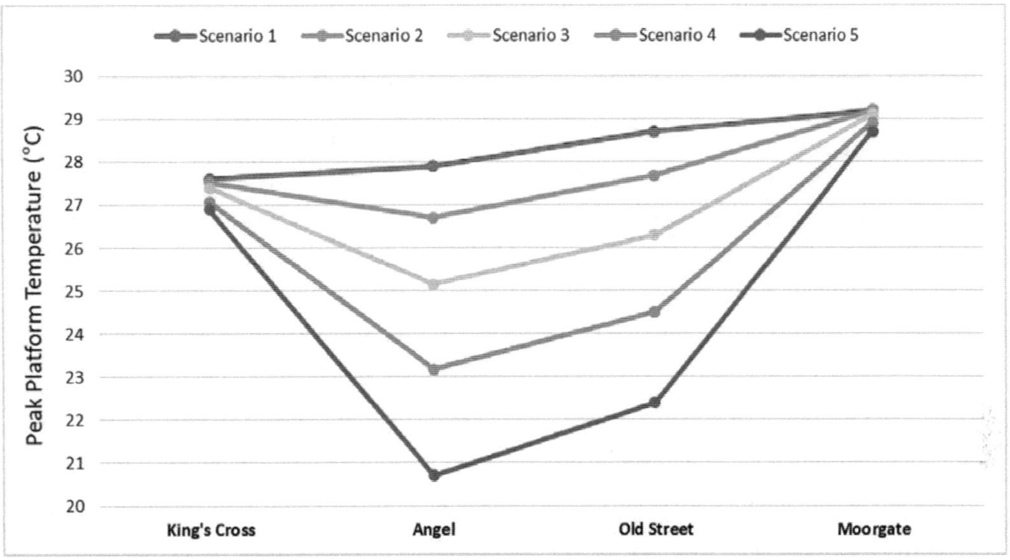

Figure 10 – Peak platform temperatures for 2030 based upon a combination of extract and supply mode SES simulations.

8. Wider opportunities for waste heat recovery

This investigation is part of the broader work from the Heating and Cooling Research Group at LSBU in the field of heat recovery, e.g. through projects LoT-NET [35] and GreenSCIES [36]. Our research group has been involved in several investigations of opportunities to recover and reuse waste and renewable heat. Underground railways represent only a small proportion of the overall low-temperature WHR potential we have estimated for the UK [37], as shown in Figure 11, which compares different unconventional heat sources that could be efficiently exploited by low-carbon DH networks using large-scale heat pumps. Although a relative small heat source, underground railways could still be used to reduce carbon emissions and energy costs locally, as shown in this study.

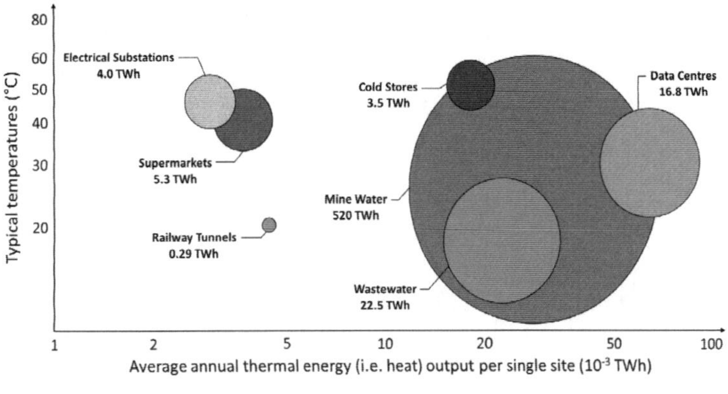

Figure 11 – Summary of total annual waste heat potential for the low-temperature heat sources considered in [37].

Overall, we have quantified the UK's waste heat potential, i.e. excluding mine water, to be 53 TWh from low-temperature sources only. This value represents 65% of the projected increase of the heat network market by 2050, which is expected to grow from meeting 14 TWh to 95 TWh of the UK's heat demand annually [38]. This indicates how waste heat and large-scale heat pumps are likely to play a key role in decarbonising the built environment as the UK moves towards net zero.

9. Conclusions

This paper presented an investigation into the potential of integrating heating and cooling as part of a heat pump system that recovers waste heat from the London Underground metro system. A mathematical model has been developed and utilised to demonstrate how the delivery of both heating and cooling through a WHR system could achieve significant benefits in terms of reducing the levelised cost of energy and maximising carbon savings. The results have indicated how the exploitation of cooling could be crucial in making WHR projects a feasible investment. The combination of both heating and cooling benefits could lead to reductions of up to 39% in the levelised cost of energy and 57% in the carbon abatement cost against ASHPs. This highlights how exploring additional benefits and the higher energy efficiencies of waste heat can compensate for its typically higher capital costs.

The provision of cooling through the WHR system also has significant impacts on the local tunnel environment, with peak temperature reductions of up to 7.2 K being estimated for adjacent stations in 2030. When operating for half the year in supply mode (i.e. when cooling is delivered), the WHR system would be able to reduce peak temperatures at adjacent stations by an average of 2.6 K compared to a do-nothing scenario. One risk identified regarding supply mode operation is the reduction of system efficiency, as lower temperature air is used as the heat source. This could increase running costs for system operators, and a balance between cooling and heating benefits must be sought. Furthermore, it is still important to highlight how the price disparity between electricity and natural gas in the UK is a major barrier to the development of WHR systems, particularly from a heating perspective. It is expected that future policy can provide much-needed support for low-carbon heat networks, allowing waste heat to play its role as an essential resource for minimising the costs of decarbonisation as the UK moves towards a clean energy future.

Acknowledgements

The authors would like to express their gratitude for the support received from Transport for London, London South Bank University, and the Engineering and Physical Science Research Council (EPSRC) through project LoT-NET - Low Temperature Heat Recovery and Distribution Network Technologies [EP/R045496/1]. The authors are also grateful to the London Borough of Islington for their support in providing valuable information for this project.

About the Lead Author

Dr Henrique Lagoeiro

Henrique is a research fellow at London South Bank University, where he investigates how the exploitation of waste heat through heat pumps and district energy can be key for decarbonising cities. Current winner of the prestigious Ted Perry Award, Henrique was also the 1st place winner at the 7th DHC+ International Awards and voted Best Young Energy Efficiency Researcher at the World Sustainable Energy Days 2022.

References

1. National Grid, 2023. Energy explained - How much of the UK's energy is renewable? Available online: https://www.nationalgrid.com/stories/energy-explained/how-much-uks-energy-renewable [Accessed 14/11/2023].
2. Department of Energy Security and Net Zero (DESNZ), 2023. UK greenhouse gas emissions, provisional figures. Available online: https://assets.publishing.service.gov.uk/media/642 4b8b83d885d000fdade9b/2022_Provisional_emissions_statistics_report.pdf [Accessed 08/12/2023].
3. Bank of England, 2022. When will inflation come down? Available online: https://www.bankofengland.co.uk/knowledgebank/will-inflation-in-the-uk-keep-rising [accessed 30/09/2022].
4. Grubb, M., Ferguson, T., Musat, A., Maximov, S., Zhang, Z., Price, J. & Drummond, P., 2022. Navigating the crises in European energy: Price Inflation, Marginal Cost Pricing, and principles for electricity market redesign in an era of low-carbon transition. UCL Institute for Sustainable Resources – Series Navigating the Energy-Climate Crises. WP #3.
5. National Energy Action, 2022. The hardest hit: Impact of the energy crisis. Available online: https://www.nea.org.uk/wp-content/uploads/2023/01/3830_NEA_Fuel-Poverty-Monitor-Report-2022_V2-1.pdf [Accessed 13/10/2023].
6. Department for Energy Security & Net Zero (DESNZ), 2022. Electricity networks strategic framework. Available online: https://www.gov.uk/government/publications/electricity-networks-strategic-framework [Accessed 14/11/2023].
7. Rosenow, J., 2022. Is heating homes with hydrogen all but a pipe dream? An evidence review. Joule 6(10), 2225-2228.
8. Rowley, P., Wilson, G. & Taylor, R., 2018. Heat decarbonisation challenges: local gas vs electricity supply. Available online: https://ukerc.ac.uk/publications/local-gas-demand-vs-electricity-supply/ [Accessed 14/11/2023].
9. Lund, H., Werner, S., Wiltshire, R., Svendsen, S., Thorsen, J. E., Hvelplund, F. & Mathiesen, B. V., 2014. 4th Generation District Heating (4GDH): Integrating smart thermal grids into future sustainable energy systems. Energy 68, 1-11.
10. Revesz, A., Jones, P., Dunham, C., Davies, G., Marques, C., Matabuena, R., Scott, J. & Maidment, G., 2020. Developing novel 5th generation district energy networks. Energy 201, 117389.
11. Lagoeiro, H., Revesz, A., Davies, G., Maidment, G., Curry, D., Faulks, G. & Murawa, M., 2019. Opportunities for Integrating Underground Railways into Low Carbon Urban Energy Networks: A Review. Applied Sciences 9(16), 3332.
12. Transport for London (TfL), 2019. Facts & Figures. Available from: https://tfl.gov.uk/corporate/about-tfl/what-we-do/london-underground/facts-and-figures [14/01/2019].
13. Duffy, S., 2018. Opportunities to Utilise Transport for London's Secondary Heat Sources. CIBSE Technical Symposium, London, UK.
14. United Nations Environment Programme (UNEP). 2022. Emissions Gap Report 2022. Available online: https://www.unep.org/events/publication-launch/emissions-gap-report-2022 [Accessed 12/12/2023].
15. Met Office. 2022. Unprecedented extreme heatwave, July 2022. Available online: https://www.metoffice.gov.uk/binaries/content/assets/metofficegovuk/pdf/weather/learn-about/uk-past-events/interesting/2022/2022_03_july_heatwave_v1.pdf [Accessed 12/12/2023].
16. Birmingham Energy Institute. 2023. The Hot Reality: Living in a +50°C World. Available online: https://www.sustainablecooling.org/wp-content/uploads/2023/11/Hot-Reality-Living-in-a-50C-World.pdf [Accessed 12/12/2023].
17. Greenham, S., Ferranti, E., Powell, R., Drayson, K. & Quinn, A., 2023. The impact of heat on London Underground infrastructure in a changing climate. Weather 78, 170-175.
18. Lagoeiro, H., Revesz, A., Davies, G., Gysin, K., Curry, D., Faulks, G., Murphy, D., Vivian, J. & Maidment, G., 2022. Investigating the opportunity for cooling the London underground through waste heat recovery. Building Services Engineering Research and Technology 43(3), 347-359.
19. Energy Systems Catapult, 2020. Thermal Energy Storage for Heat Networks. Available online: https://es.catapult.org.uk/report/storage-and-flexibility-thermal-energy-storage-for-heat-networks/ [Accessed 16/11/2023].

20. Chartered Institution of Building Services Engineers. 2020. CP1 – Heat Networks Code of Practice for the UK.

21. Heinonen, E. W., Wildin, M. W., Beall, A. N. & Tapscott, R. E., 1997. Assessment of Antifreeze Solutions for Ground-Source Heat Pump Systems. ASHRAE Transactions 103(2), 747-56.

22. Lagoeiro, H., Revesz, A., Davies, G., Gysin, K., Curry, D., Faulks, G., Murphy, D., Vivian, J. & Maidment, G. 2022. Integrating Waste Heat Recovery from Railway Tunnels into Flexible Heat Networks. Proceedings of the ASHRAE Winter Conference, USA, 29 Jan – 2 Feb, 2022.

23. Ibrahim, N. & Kennedy, C. 2016. A Methodology for Constructing Marginal Abatement Cost Curves for Climate Action in Cities. Energies 9(4): 227.

24. European Commission. 2023. Reference and discount rates (in %) since 01.08.1997. Available online: https://competition-policy.ec.europa.eu/state-aid/legislation/reference-discount-rates-and-recovery-interest-rates/reference-and-discount-rates_en [Accessed 31 July 2023].

25. Danish Energy Agency, 2023. Technology Data. Available online: https://ens.dk/en/our-services/projections-and-models/technology-data [Accessed 23/11/2023].

26. Energy Saving Trust, 2013. The heat is on: heat pump field trials: phase 2. Available online: https://www.energysavingtrust.org.uk/sites/default/files/reports/TheHeatisOnweb%281%29.pdf [Accessed 13/04/2021].

27. Department for Energy Security & Net Zero (DESNZ), 2023. Quarterly Energy Prices. Available online: https://www.gov.uk/government/collections/quarterly-energy-prices [Accessed 18/12/2023].

28. Hamilton, J., 2022. Institute of Refrigeration SIRACH Event: Decarbonising heat using secondary sources and Government policy - Waste Heat. Available online: https:// ior.org.uk/public/downloads/HD2mP/Dr-Joel-Hamilton-June-22.pdf [Accessed 01/12/2023].

29. Revesz, A., Dunham, C., Jones, P., Bond, C., Fenner, R., Mody, S., Nijjhar, R., Marques, C. & Maidment, G. 2022. A holistic design approach for 5th generation smart local energy systems: Project GreenSCIES. Energy 242, 122885.

30. Department for Energy Security & Net Zero (DESNZ), 2023. Green Book supplementary guidance: valuation of energy use and greenhouse gas emissions for appraisal. Available online: https://www.gov.uk/government/publications/valuation-of-energy-use-and-greenhouse-gas-emissions-for-appraisal [Accessed 30/07/2023].

31. Department for Energy Security & Net Zero (DESNZ), 2022. Green Heat Network Development Fund Round 1: Guidance for Applicants, March 2022, Available online: https:// www.gov.uk/government/publications/green-heat-network-fund-ghnf [Accessed 15/11/2022].

32. Thompson, J.A., Maidment, G.G. & Missenden, J.F., 2006. Modelling low-energy cooling strategies for underground railways. Applied Energy 83, 1152-62.

33. U.S. Department of Transportation, 2002. Subway Environmental Design Handbook, Volume II. Subway Environmental Simulation Computer Program, Version 4.1.

34. Wen, Y., Leng, J., Shen, X., Han, G., Lijun, S. & Yu, F., 2020. Environmental and Health Effects of Ventilation in Subway Stations: A Literature Review. International Journal of Environmental Research and Public Health 17, 1084.

35. Low Temperature Heat Recovery and Distribution Network Technologies (LoT-NET), 2023. Project website. Available online: https://lot-net.org/ [Accessed 19/12/2023].

36. Green Smart Community Integrated Energy Systems (GreenSCIES), 2023. A Centre of Excellence in SLES. Available online: https://greenscies.com/ [Accessed 19/12/2023].

37. Davies, G., Lagoeiro, H., Turnell, H., Wegner, M., Foster, A., Evans, J., Revesz, A., Leiper, A., Smyth, K., Hamilton, J., Cooke, H. & Maidment, G., 2023.Evaluation of low temperature waste heat as a low carbon heat resource in the UK. Applied Thermal Engineering 235, 121283.

Waste Heat from the London Underground

Dermot Cotter – You're the winner of the 2022/2023 Ted Perry award and you covered two areas of your PHD during this talk. What did you find were the biggest challenges during your PHD on those topics? And also what was the easiest aspect to research?

Henrique Lagoiero – Data was quite challenging because we were trying to validate the model while the system was being commissioned so having access to data was a challenge. Having enough data to validate the system and see that the heat pump was working as it should. That was the main challenge because these were things outside my control. The easiest thing was when I decided what I actually wanted to investigate. That gave me peace of mind because then I could do my research and try to develop the model.

DC – It sounds as if it's not the technical bits that are challenging, it's all the bits around it. When I look at the London Underground in simplistic terms, there's a train running around and electricity is feeding it. I assume that with braking and motor power that any electricity that goes goes in comes out as heat eventually. Am I correct?

HL – There are some regenerative braking systems so some energy is fed back to power lights and doors but most energy ends up being degraded and released inside the tunnels as waste heat.

DC – You gave an example of how you are putting air down to cool the tunnels. What would happen if there was a train strike for a week. Would the tunnels cool down very quickly or is there a store of residual heat that would take a long time to come out or would you lose your heating system for local buildings quite quickly?

HL – The main source of heat is the running of the trains. We didn't have much chance to look at the changes overnight when the trains stopped running but I suspect that in a strike the tunnels would cool quite quickly. However the soil around the tunnels is saturated with the heat and that would be released back into the tunnels. We found from our data that the temperature in the tunnels was quite actually quite stable.

DC – You mentioned having storage of the heat. Is that a necessity when you already have the storage in the ground or are you needing more heat than the heat pump can deliver?

HL – There is the energy there to generate heat at different times but I think the storage can be efficient to match when it is least costly to generate electricity, which might not be when you have the lowest COP. Additional storage is beneficial so you're not only relying on the heat that is there but you can also respond to the market. We did study this in the PhD and it was found that storage can really help to reduce the operational costs of the system.

DC – I imagine the underground is a very dirty environment that is hard to keep clean. Are there any issues with coils getting dirty or blocked and reducing the performance?

HL – It can clog easily and there were some issues. There was some work carried out before my research looking at installing different types of coils, with filters and without filters and different size fins, looking at the optimum design for less cleaning and replacement of filters. It was realised that at 4 fins per inch that there would be some build up but that after a certain point there would not be any further build up because the gap was large enough to allow particles through without clogging. It would still need regular cleaning to make sure the coils don't get clogged and that can be a challenge.

DC – You talked about Bun Hill 1 and Bun Hill 2. Are the heat networks now connected to each other or are they still separate?

HL – They are connected. At Bun Hill 1 the operating temperature was 95 and the return temperature was 75. They reduced the whole temperature to be able to incorporate the heat pump. It is now all connected so it is essentially one network with two energy centres. Bun Hill 1 has a CHP and Bun Hill 2 has a heat pump.

DC – How does it balance out which it is using? Are both energy centres always running flat out or does one take over from the other as needed?

HL – We didn't get to see the heat pump in full operation but the strategy is to prioritise the heat pump and run it as much as we can and then use the CHP for peak demand. The original communal gas boilers were also kept connected to the system in case there was any issue or a top up was needed. The hierarchy is to run the heat pump as the most carbon efficient system and to use the CHP to meet additional demand.

DC – When looking at investments and long-term costs. Looking at efficiency is quite easy, it is the capital costs that are awkward. How much assistance did you need to get for the capital costs of all this?

HL – It was really challenging, especially in the beginning. The previous project we did funded by BEIS and ARUP really helped because we had expertise from the government and ARUP to support with some benchmarks and capital costs. It is challenging but there is data out there. At the beginning we only focussed on operational costs because it was difficult to get data on capital costs but as the project developed we were able to get some updated estimates to use for the model.

DC – I assume they wanted to know the financial cost before you could even look at the technical aspect of the project. Last question. What is the COP in the winter compared with the summer?

HL – The COP in the winter reduced quite significantly. It can down to 2.5, it really depends what the temperature is and you can get really high COPs in the summer. We could get COP of around 3.4, even 4.5 or 5. It really depends what heat source you're using.

IOR.org.uk
REFRIGERATION AIR CONDITIONING HEAT PUMPS

The knowledge hub for refrigeration, air conditioning and heat pumps

The Challenges of Heat Pumps for Domestic Heating

Christopher Parker AMInstR

What you will learn

- The paper summarises the results of an investigation and modelling of the cost and energy supply and carbon emission consequences from electrification of domestic heating using electric heat pumps to meet the UK 2050 net zero target.
- A key component of the government strategy for the transition to heat pumps is the thermal insulation upgrade of legacy housing to current standards.
- The modelling predicts a total cost of £306bn to switch domestic heating to electric heat pumps by 2050 and will require around a 2.4x increase in electricity generation capacity and a large deployment of carbon capture storage at power stations if an increase in UK carbon emissions is to be avoided.
- A key challenge for the industry is the number of domestic heat pump installations that will have to be delivered to meet the 2050 target. The modelling shows a rapid ramp up in annual installation numbers, rising to around 1 million in 2030 and around 2 million a year by the mid 2030's.

Abstract

The financial and electricity burden caused by the electrification of domestic heating has been investigated and presented in this paper. The overall cost to achieve legacy housing improvement and widespread ASHP adoption is approximately £309 billion pounds between 2023 and 2050. The financial burden will vary for each household, but most of the burden falls on legacy housing from pre-1919 and gradually reduces as property age reduces to 1981-1990 period. Multiple fund schemes financed by UK Government and local authorities will ease the burden, but they are targeting low-income housing representing 21% of the domestic market and are only scheduled to operate until 2026. Further and clear financial support is required to encourage insulation refits and heat pump adoption.

The increased electricity demand requires an installed plant capacity of 248 GW by 2050 with the generation split of 39% renewables, 32% CCCT with carbon capture storage and 29% nuclear, a 137% increase from the existing capacity. Despite no demonstrable case study for the scale needed, a cornerstone of the UK Government strategy is the deployment of carbon capture storage (CCS) technology as a reliance on natural gas is retained for secure grid flexibility while fulfilling the definition of 'Net Zero' (1). Without CCS, the total consumption of natural gas increases in an electrified society compared to retaining domestic natural gas boilers by 2.9%, thereby increasing annual emissions. If a completely renewable UK society was targeted by 2050 an installed capacity of 435 GW would be required, an increase of ten times the current renewable capacity.

Introduction

The UK Government is currently committed to reducing the net carbon account in 2050 to 100% lower than the 1990 baseline (2). To achieve this goal, every sector must seek methods to reduce their carbon emissions. Heating domestic properties accounted for 17% (77.3 $MtCO_2e$) of the UK's territorial carbon emissions in 2019 (3). Current strategies proposed by policy makers and industry experts is to move domestic heating technology from natural gas boilers to a mixture of heat pumps, hydrogen boilers and communal/district heating systems, depending on the unique property challenges (4). This paper proposes the challenge of nationwide conversion of 28,679,806 (as of 2022) domestic properties (5), aspired to increase to 28,802,646 by 2025 and 34,020,553 by 2050, to an electrified heat pump solution. The challenges focus on the financial burden to prepare the substantial legacy housing stock for heat pump technology and the growing impact on the UK's electricity supply in line with current Government strategy, ambitions and timelines.

How Heat Pumps Work in A Home

A heat pump is a mechanical device that transfers heat from a cool sink to a warm sink using the refrigeration cycle. The type of cool sink often describes the type of heat pump. Air source heat pumps transfer heat from mechanically motivated air external to the control space. Ground source heat pumps transfer heat from the soil or ground water and water source heat pumps transfer heat from an open body of water. There are other types of heat pumps available, but the three mentioned are the most accessible in the UK domestic market.

A heat pump system requires a hot water storage cylinder as the technology is not efficient at generating direct hot water but rather maintain a provision of instant hot water (4). Since central heating became mandatory for newly constructed housing in 1970, most domestic properties have a refurbished central heating system from within the past 20 years. The original systems were open vented with hot water cylinder storage but have gradually been replaced with combination boilers, eliminating the need for water storage space.

As of 2021, there were 15 million combination boiler systems and 9 million boiler systems with a hot water storage component. The original mounting space for the 15 million storage cylinders made redundant by past refurbishments have likely been repurposed for alternative use (4).

The installation of heat pumps will require reinstallation of a water storage cylinder which may necessitate sacrificial dwelling space to accommodate the vessel as original mountings may have been re-purposed, and dwellings constructed over the past two decades do not have accommodation for a storage vessel (4). Recent guidance has suggested a combined approach is required to make a property 'low-carbon' ready by ensuring the building fabric is of suitable insulation level, the heat emitters and pipework are of appropriate size and a hot water storage space is made available (4). These, combined with the change of technology from boiler to heat pump increases the financial burden of the heat pump installation. As of 2022, 97% of domestic heating demand was supplied by a fossil fuel energy source and approximately 83% was supplied by natural gas fired boilers (6).

UK Housing

The UK has the oldest housing stock in Europe and most likely the world (6), consequently, the energy efficiency

Table 1 - Number of properties in the UK by age, 2017 data (6). (1,000s)

Dwelling Period	Nation				UK Total
	England	Scotland	Wales	Northern Ireland	
Pre 1919	4,972	467	351	82	5,871
1919-1944	3,793	291	133	68	4,284
1945-1964	4,582	544	219	126	5,472
1965-1980	4,689	515	304	189	5,698
1981-1990	1,895	194	99	99	2,287
Post 1990	4,019	452	235	216	4,923

of UK housing is amongst the worst in Europe (7). There are several historical events that have led to a relatively even split of dwelling periods in the UK as listed in Table 1.

The presence of properties from various periods introduces unique challenges to replace fossil fuel heating with electrified heat pumps across the country. The Government's current strategy is to improve the thermal energy efficiency of the existing housing stock in preparation for heat pump central heating (3).

Modern properties constructed with the latest standards and guidelines require little modification to reduce energy consumption whereas, the further back in time the property was constructed, typically more work is required. Prior to 1919, most housing was of solid wall construction, with cavity walls being introduced during the inter-war period before becoming standard after The Second World War. While cavity wall construction became normal practice, installing insulation within the cavity wall was not introduced until the 1970s and was not a compulsory requirement until 1990 (8).

This timeline leads to the worst-case scenario of 23,612,000 dwellings with no wall insulation, however, Government grants since the 1990s has made retrofitting cavity wall insulations a cost effective and common energy efficiency measure (6). Solid wall insulation is available but not widely undertaken due to a mixture of higher installation costs from constructing either an internal or external cavity wall prior to insulation and difficulty in properly detailing the most appropriate design for construction without introducing condensation

Table 2 – Number of UK Dwellings with and without wall insulation (6) (1,000s).

Dwelling Wall Construction	Nation				UK Total
	England	Scotland	Wales	N. Ireland	
Cavity insulated	11,157	1,363	636	570	13,726
Cavity uninsulated	5,242	457	298	66	6,063
All cavity walls	16,399	1,820	934	636	19,789
Solid with insulation	694	115	71	11	891
Solid uninsulated	6,301	529	307	107	7,244
All solid walls	6,996	644	378	118	8,136

(8). Table 2 lists the volume of homes with wall insulation by construction type in 2017 (6).

Using the data and knowledge of the construction timeline, an estimation of the volume of insulated walls and uninsulated walls by dwelling age can be made. Table 3 list the approximate split of wall insulation.

Table 3 - Approximate number of households with wall insulation by construction type and dwelling period (1,000s).

Dwelling Period	Solid Walls	Cavity Walls	Estimate With Insulation	Estimate Solid Walls Without Insulation	Estimate Cavity Walls Without Insulation
Pre 1919	5,871	-	696	5,175	-
1919-1944	2,265	2,019	1,457	1,996	830
1945-1964	-	5,472	3,222	-	2,250
1965-1980	-	5,698	3,355	-	2,343
1981-1990	-	2,287	1,347	-	940
Post 1990	-	5,678	5,678	-	-
Totals	8,136	21,154	15,755	7,171	6,363

There are approximately 13.5 million homes in the UK without wall insulation. Further incentives have delivered approximately 17 million insulated loft space across all property types according to the latest standards (270mm thick insulation) up to December 2022 (9).

Table 4 - ECO+ consultation estimated insulation cost (10).

Property Type	Loft insulation	High-cost cavity wall	Solid wall insulation
Terraced	£890	£8,708	£22,167
Semi-detached	£950	£9,533	£24,267
Detached	£1,300	£12,742	£32,433
Bungalow	£970	£8,617	£21,933
Flat	£890	£7,333	£18,667
Annual Thermal Energy Reduction kWh (11)	550	1,683	4,107

Current UK Government consultation has further extended insulation grants for low income and low EPC rated housing between 2023 and 2026 through the ECO+ scheme and provides an estimate cost and scale to install insulation initiatives across various properties (10), see Table 4.

Based on the UK Government's estimated figures in Table 4 and the published housing statistics from the last National Housing Survey listed in Table 3 (6), the following Figure 1 illustrates the capital investment required insulate the remaining UK housing stock to the current building standards.

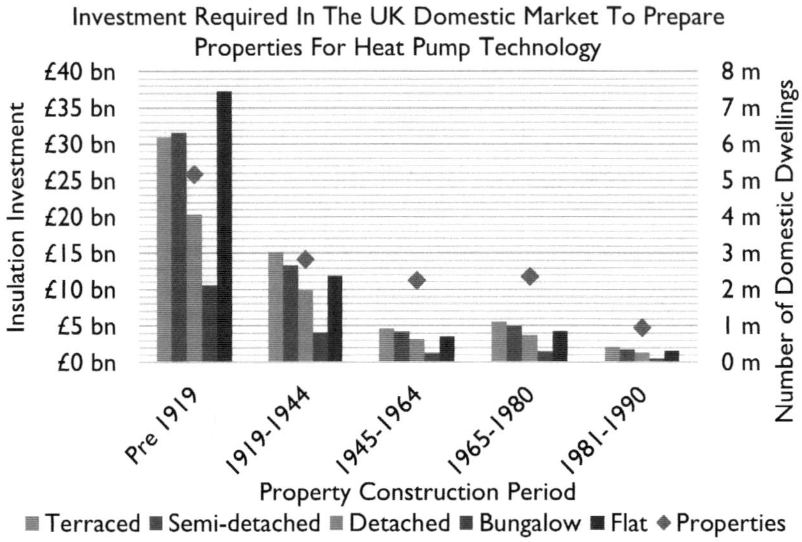

Figure 1 - Insulation investment required to prepare the UK housing stock for efficient heat pump installations by property type and period.

The approximate total capital investment required is £229 billion pounds to modernise the remaining legacy housing stock thermal efficiency levels to current building standards. Of the total cost, 57% is for pre-1919, 24% for 1919-1944, 7% for 1945-1964, 9% for 1965-1980 and 3% for 1981-1990 housing.

Over the years to 2050, the number of old housing stock will likely remain the same as it currently stands due to low replacement rates in the UK and the continuing desirability for older homes (6). Therefore, it is reasonable to expect the modernisation investment to be spread over the period.

The UK Government's Heat Strategy includes seven key ambitions related to heat pumps on the journey to 2050 (3).

- By 2025, there is an ambition for heat pumps cost to reduce 25% to 50% of 2021 prices (£7,000 to £14,000).
- By 2025, natural gas boilers cannot be installed in new UK housing.
- By 2028, 600,000 heat pumps should be installed annually, 50% of which are manufactured in the UK.
- By 2030, 1,000,000 heat pumps should be installed annually.
- By 2030, heat pump costs should reach parity with natural gas boilers.
- By 2035, natural gas boilers cannot be installed in legacy UK housing.
- By 2050, the UK should achieve Net Zero.

Assuming the ambitions are achieved and the following expectations are made, a model of domestic heat pump growth, investment and fossil fuel decline to 2050 can be made.

- UK population growth continues at a rate equal to the average growth rate between 2011 and 2020 (+0.67%) (5)
- Natural gas boiler cost £2,000.

- ASHP initially cost £10,000.
- Annual new housing growth ensures an average of 2.4 persons per household is maintained.
- All new housing from 2023 have Air Source Heat Pumps (ASHPs) installed during construction.
- ASHPs installed annually between 2023 (200,000), 2028 (600,000), 2030 (1,000,000) and 2035 (replacement gas boiler ban) increase linearly.
- Natural Gas boilers have a life expectancy of 15 years.
- ASHPs have a life expectancy of 20 years.
- Heat pump retrofit installations include the cost to insulate legaçy housing to modern standards.
- Annual retrofit installations are equally split between dwelling type and period construction.

Figure 2 illustrates the modelled change of installed system types and the investment fluctuations between 2023 and 2050 for the idealised Government strategy.

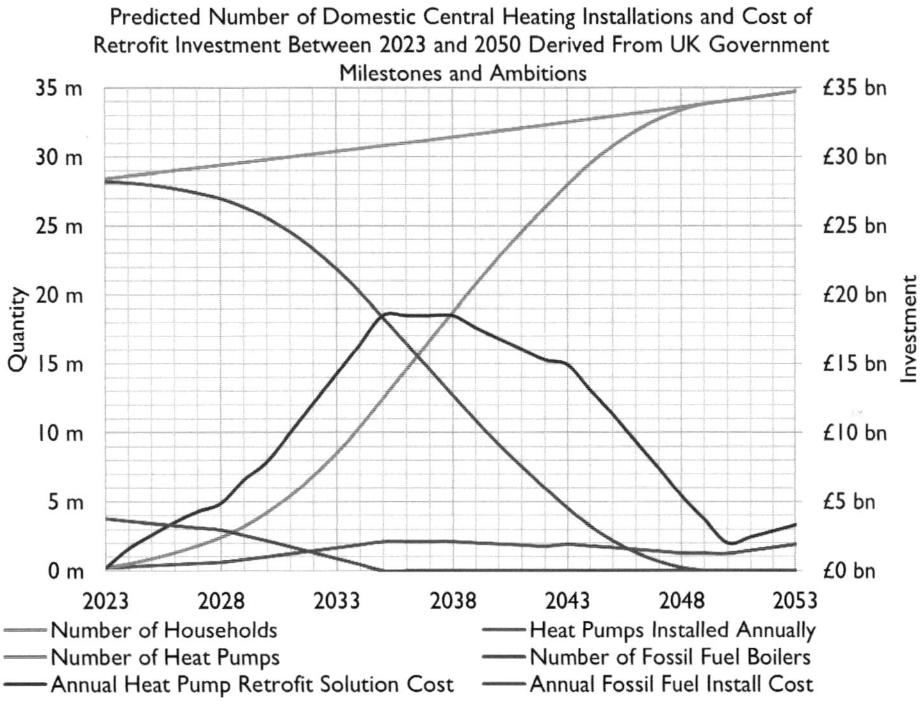

Figure 2 - Predicted growth of domestic heat pump installations and decline of natural gas boiler central heating.

The number of heat pumps installed increases sinusoidally to equal the total housing stock by 2050 while natural gas boilers reduce inversely to 0 by 2050. A total of £85.7 billion pounds is required to install heat pumps into every household between 2023 and 2050 after the insulation modification. The annual heat pump retrofit investment increases sharply from approximately £1.6 billion to a peak of £18.5 billion in 2035 before gradually decreasing to £2 billion in 2050. The cause of the increase is the targeted heat pump installations of the Government ambitions increasing the annual number of targeted legacy housing which are further invested in to increase thermal efficiency. The gradual decline of annual investment after 2038 is due to the reduced number of boiler retrofits as the 2023 installations reach their end of life, and the volume installed is reduced by the heat pumps installed that year.

The decline is continuous until 2043 where the first-generation heat pumps installed in 2023 come to their end of life and require replacement, but the insulation works remain intact. The annual investment continues to decrease until 2050 when all legacy housing energy efficiency has been improved to current building standards and thereafter, heat pump installations are the gradual end of life replacements from 20-year previous retrofits or new build construction.

This model does have limitations, mainly:

- The assumption that all domestic heating will be supplied by ASHP in 2050 is highly unlikely due to continuing research and funding into alternative forms such as ground source heat pumps, district heating and hydrogen boilers likely splitting the percentage of solutions by 2050. Although, no definitive strategy has been outlined.
- Individual flats have been considered to receive their own ASHP solution when communal heating systems would be more likely for a flat block. However, there is limited data available to determine the spread of flats across block sizes in the UK housing survey, restricting the model to assume a flat-by-flat solution.
- Insulation upgrade figures are fixed according to the Government ECO+ assumptions for each housing type whereas the cost would vary by building size, planning limitations and competition; as a highly valuable market, more installers and manufacturers would likely establish themselves to supply a demanding market.
- Value changes in the sterling pound have not been considered as there is no viable method to predict currency change.
- An even split of dwelling types and period has been considered for the annual retrofit as it is unknown whether certain demographics would respond to heat pump installations quicker than others.

While the limitations are stated, the modelled Government strategy clearly illustrates a strong financial burden on the domestic market, over five times greater than the natural gas boiler replacement by 2035. Further Government proposals suggest providing financial grants for low-income housing through various schemes (3) but, as low-income is generally classified as an annual income below 60% of the median income of the UK population (12) this represents approximately 6 million or 21% of all UK households (13) leaving the remainder to finance the insulation modification themselves with potential support from local authority grants.

UK Energy Market Since 1970

The UK's electricity generators have changed significantly since 1970. In 1970 coal was the prominent source of heat to generate electricity by the Rankine Cycle. Following coal, the next significant fuel sources were oil, nuclear, hydro and natural gas. Alternative fuels such as biofuels and renewable generators were available but did not generate electricity in significant numbers (14). This split of significant fuel sources remained consistent throughout the 1970s and 1980s despite the declining coal mining industry in the UK. It was not until 1992 that generation from coal was gradually replaced with Natural Gas following the sudden expansion of gas fired generation in what was known as the 'Dash for Gas' (15). Combined Cycle Gas Turbines (CCCT) continued to be rapidly expanded to dominate the energy generation market until 2013 where they held their peak split of the power capacity make-up and have since maintained a consistent capacity value (14). Renewable energy generators had also been gradually expanding alongside the 'Dash for Gas' although not at the same magnitude. It was not until 2010 that renewable generation started rapidly expanding to the portion seen in 2022 while coal plant capacity has receded to less than hydro capacity.

As of the 2022 data, renewable generators are continuing to expand and are not showing signs of plateauing (1). The following Figure 3 illustrates the electricity generator split by energy source in 1970 and 2022.

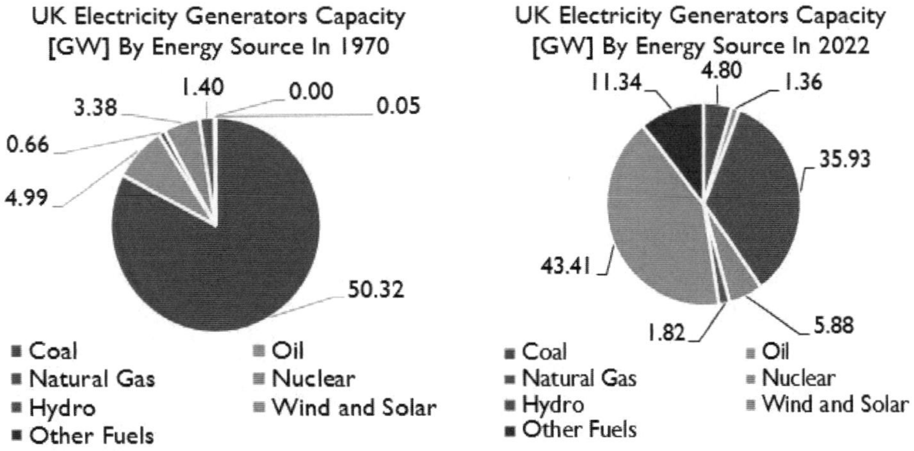

Figure 3 - UK electricity generators installed capacity [GW] by energy source in 1970 and 2022 (14)

Several other features of the UK electricity supply have changed since 1970 and have been listed in the following Table 5.

Table 5 - Electricity generation changes between 1970 and 2022.

Year	1970	2022
Installed Capacity	61.56 GW	104.55 GW
Load Factor	43.2%	33.9%
Domestic Use	33%	34%
Carbon Factor	879 gCO_{2e}/kWh	207 gCO_{2e}/kWh

The installed capacity represents the sum power generation of all energy sources and has increased over the period due to population growth. The load factor represents the percentage total of energy supplied from the potential installation capacity and had reduced due to the increased reliance on renewable resources that have a limited operating period compared to fuel source generators. Domestic use is the percentage of total electricity supply that is used by domestic households and has remained relatively consistent by only increasing from 33% to 34%. The carbon factor is the representative greenhouse gas (GHG) emissions released by the total electricity generation for the year from the fuel input supplied. The carbon factor has decreased due to a reduced reliance on coal, which is the most polluting fuel source in terms of direct GHG emissions and further increasing reliance on renewable resources (14) (16) (17).

The data available shows how electricity generation has changed over 50 years from 1970 to 2022 but also emphasises the challenge to achieve a net zero electricity supply by 2050. Using the heat pump installation model developed against the UK Government's own timeline and ambitions, the future impact on the UK's energy supply can be predicted and used to target increased installation capacity.

A current unknown is how the UK power generator capacity will be split by energy source in 2050. There is no defined strategy or ambition agreed for 2050 and the UK Government are consulting experts in various committees to learn what options are available.

One committee paper delivered to the BEIS Select Committee on 9 January 2023 suggests a split between 39% renewables, 32% CCCT with carbon capture storage and 29% nuclear (18) which closely corresponds with the current published Government material (1).

Several assumptions must be made in addition to those listed for the heat pump installation model:

- Annual domestic energy use for a society that retains natural gas for domestic heating reduces from an initial values of 11,309 kWh natural gas and 4,053 kWh electricity per household in line with insulation installation following the ASHP installation model. (14) (17)
- The split of domestic natural gas use for a society that retains natural gas for domestic heating remains 36% of the total UK supply each year, representing a proportional growth in commercial and industrial markets. (14) (17)
- The split of domestic electricity use for a society that retains natural gas for domestic heating remains 34% of the total UK supply each year, continuing the represented proportional growth in commercial and industrial markets. (14) (17)
- The transition of power generator split from the 2022 values 43%/34%/6% for renewables, CCCT and nuclear is linear to the select committee paper proposal values. (18)
- CCCT thermal efficiency remain fixed at 46%. (14) (17)
- The load factor of renewables, CCCTs and nuclear remain fixed at 24%, 37% and 71% respectively. (14) (17)
- For the electrified future society where, electrified transport is achieved by 2050, the travel miles per household is 11,290 per year (19).
- Electric vehicles are capable of 0.33 miles per kWh.
- The ASHP installed maintain a seasonal coefficient of performance of 2.5.

The following Figure 4 illustrates the changes in energy consumption for three development strategies. The first is a UK society that does not electrify domestic heating but rather retains the use of natural gas boilers. While this is contradictory to the UK's net zero ambitions, it establishes a baseline trajectory. The second considers the gradual installation of ASHP in all domestic properties following the model previously described and discussed. The third considers the ASHP installation and the gradual transition of domestic transport from internal combustion engines to electric vehicles, illustrating the impact of a completely electrified UK domestic market from 2022 to 2060.

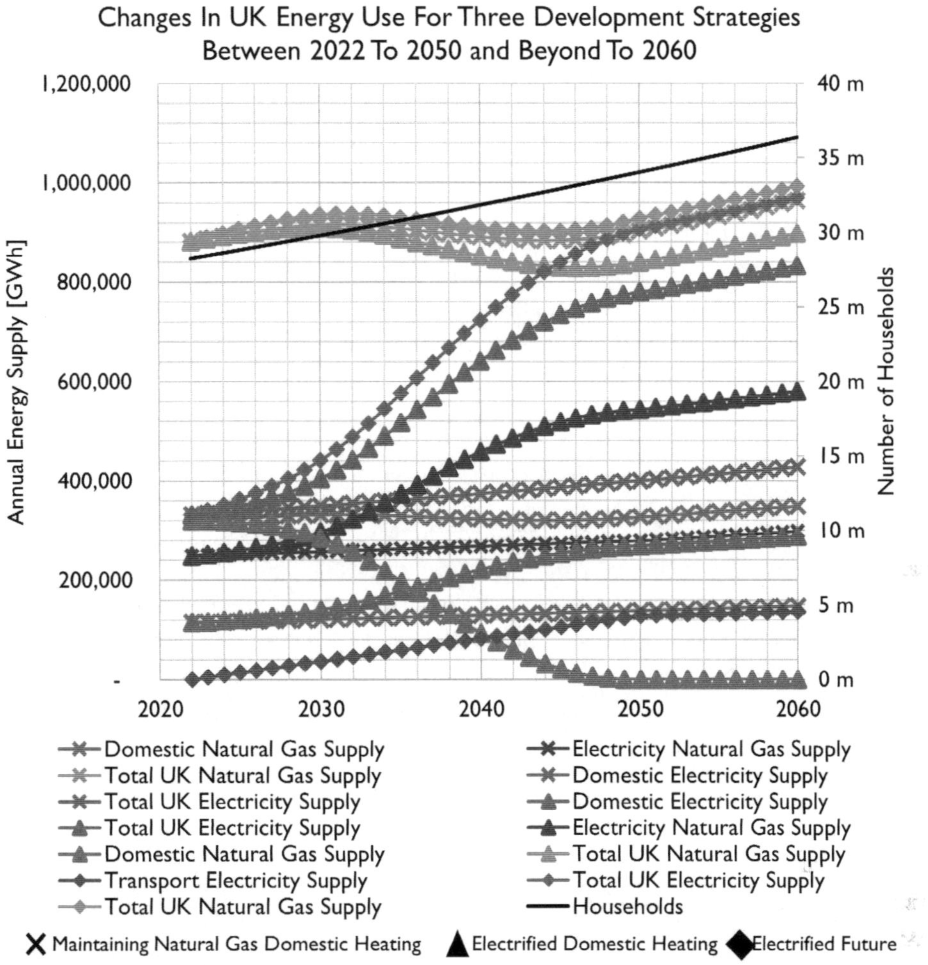

Figure 4 - Changes in UK Energy Use For Three Development Strategies Between 2022 To 2050 and Beyond To 2060

The initial strategy where natural gas is not removed from homes demonstrates a gradual increase in natural gas and electricity consumption proportional to household growth. An expected trend but establishes a benchmark for alternative strategies to measure from. By 2050, the UK electricity supply has increased by 68,058 GWh (20%) from 2022 requiring an installed plant capacity of 109 GW. To accommodate domestic heating and electricity generation, natural gas consumption is 902 TWh by 2050.

The second strategy that only considers the deployment of ASHP in the timeframe of the model discussed previously illustrates the reduction of domestic natural gas to zero and increase of electricity consumption until 2050 when the last natural gas boilers are decommissioned. After 2050 the electricity consumption follows a linear growth proportional to increasing households. Following the BEIS committee power generation split proposal, natural gas is retained in the UK for power generation and alternative uses beyond the domestic market. By 2050, the UK electricity supply has increased by 447,185 GWh (135%) from 2022 and 379,127 GWh (95%) from the benchmark 2050, requiring an installed plant capacity of 213 GW. To accommodate electricity generation, natural gas consumption is 839,386 TWh by 2050, a decrease of 4.8% from the benchmark and 6.9% from 2022.

The third strategy considers the deployment of ASHP like the second and the electrification of transport, illustrating the total impact of a completely electrified domestic market. The energy trends are like the second strategy; however, the magnitudes are larger. By 2050, the UK electricity supply has increased by 575,218 GWh (173%) from 2022 and 507,160 GWh (127%) from the benchmark 2050, requiring an installed plant capacity of 248 GW. To accommodate electricity generation and other uses, natural gas consumption is 928,572 TWh by 2050, an increase of 2.9% from the benchmark and 5.3% from 2022.

This model does have limitations in addition to those stated previously. The model assumes natural gas and electricity use outside of the domestic market remains consistent with assumed population growth representing no initiatives to increase energy efficiency or reduce natural gas beyond domestic dwellings. In the model, the thermal efficiency of CCGT remains fixed at 46% on the journey to 2050 but will likely increase due to improved technical and control advancements.

The additional demand on the UK electricity generation during the journey to 2050 is a challenge introduced by the electrification of domestic dwellings. Even if the current expectations for plant capacity and generation split can be met, the increased total consumption of natural gas cannot be ignored but acknowledged as a direct cause of increased electricity demand and energy security.

If a completely renewable energy society was possible by 2050, modifications to the model provide an expectation of an installed capacity requirement of 435 GW, a 187 GW increase from the third strategy. The large capacity requirement is directly caused by the relatively low load factor associated with renewable technology, representing the limitations of extracting wind power and solar power when the sources are available.

Conclusion

The domestic financial and electricity burden caused by the electrification of domestic heating has been investigated and presented in this paper. The cost of electricity generator and grid upgrades and workforce recruitment and training needed to deliver electrification of domestic heating were not considered by the modelling and are outside the scope of this paper.

The transition of domestic heating to heat pump technology is a key strategy to reduce the UK's annual emissions by removing the natural gas combustion in households and relying on a 'low carbon' electricity supply with the ambition to achieve a net zero by 2050. The current approach is to ensure legacy housing is modernised to existing insulation standards and improve domestic building energy efficiency. The overall cost to achieve legacy housing improvement and widespread ASHP adoption is approximately £309 billion pounds between 2023 and 2050. The financial burden will vary for each household, but most of the burden falls on legacy housing from pre-1919 and gradually reduces as property age reduces to 1981-1990 period. Multiple fund schemes financed by UK Government and local authorities will ease the burden, but they are targeting low-income housing representing 21% of the domestic market and are only scheduled to operate until 2026. Further and clear financial support is required to encourage insulation refits and heat pump adoption.

The additional demand on the electricity supply has been demonstrated for a growing UK population transitioning domestic heating to electricity as well as the parallel transition from internal combustion engines to electric vehicles, both of which are key strategies to achieve net zero by 2050. The demand requires an installed plant capacity of 248 GW with the generation split of 39% renewables, 32% CCCT with carbon capture storage and 29% nuclear, a 137% increase from the existing capacity. A limitation of the current grid expansion strategy is the lack of clear direction and ambition.

The UK Government are in the process of consulting select committees for advice and guidance with various proposals for future energy demand, the proposed split of generators being taken from one of the committees. This paper provides a value for the burden caused by electrifying the domestic market. Despite no demonstrable case study for the scale needed, a cornerstone of the UK Government strategy is the deployment of carbon capture storage (CCS) technology as a reliance on natural gas is retained for secure grid flexibility while fulfilling the definition of 'Net Zero' (1). Without CCS, the total consumption of natural gas increases in an electrified society compared to retaining domestic natural gas boilers by 2.9%, thereby increasing annual emissions. If a completely renewable UK society was targeted by 2050, an installed capacity of 435 GW would be required, an increase of ten times the current renewable capacity.

About the author

Chris Parker graduated from Liverpool John Moores University with a first-class honours degree in mechanical engineering, and early experience working as an engineering consultant. After two years of advancing his career, Chris achieved Chartered Engineer status at the age of 26. Using his knowledge and experience, Chris has supported several projects that require an in depth understanding of the thermodynamic and fluid mechanic phenomena that occur in complex mechanical systems. Chris maintains a strong comprehension of the wider landscape, seeking to learn the wider issues compound existing challenges, influence decision factors and impact engineering solutions. To this end, Chris has delved into research regarding the UK's challenges to decarbonise the national energy demands. Several opportunities are available, and many hurdles stand in the way of widespread alternative technology adoption. Chris currently works for WAVE.

Presented before the IOR on 7 March 2024 and amended by the author on 23 April 2024

References

1. His Majesty's Government. Powering Up Britain Energy Security Plan. london : His Majesty's Stationary Office, 2023. ISBN 978-1-5286-4018-3.

2. Climate Change Act 2008. Climate Change Act 2008. London : Her Majesty's Stationary Office, 2019.

3. Kwarteng, Kwasi. Heat and Buildings Strategy. London : Her Majesty's Stationary Office, 2021.

4. Heating and Hot Water Industry Council. Heating Up to Net Zero. Kenilworth : Heating and Hot Water Industry Council, 2021.

5. Department for Business, Energy & Industrial Strategy. Energy Consumption in the UK (ECUK): Energy Intensity Tables. Digest of United Kingdom Energy Statistics (DUKES). Westminster, Greater London, United Kingdom of Great Britain and Northern Ireland : Department for Business, Energy & Industrial Strategy, 29 July 2022.

6. Piddington, Justine, et al. The Housing Stock of The United Kingdom. Watford : BRE Trust, 2020.

7. Harris, Ben and Walker, Alan. UK Parliament Post Heat Pumps. London : His Majesty's Stationary Office, 2023.

8. Glew, David, et al. Demonstration of Energy Efficiency Potential. London : Department for Business, Energy & Industrial Strategy, 2021.

9. Department for Energy Security & Net Zero. Household Energy Efficiency Data to December 2022. London : His Majesty's Stationary Office, 2023.

10. Department for Business, Energy & Industrial Strategy. Energy Company Obligation ECO+: 2023 - 2026. London : His Majesty's Stationary Office, 2022.

11. Department for Energy Security & Net Zero. Household Energy Efficiency Statistics. London : Department for Energy Security & Net Zero, 2023.

12. Citizens Advice. Briefing: Local Citizens Advice Income Profile. London : Citizens Advice, 2015.

13. Office for National Statistics. National Statistics. The effects of taxes and benefits on household income: financial year ending 2021. Newport : Office for National Statistics, 2022.

14. Department for Business, Energy and Industrial Strategy. Estimated electricity capacity and generation timeseries between 1920 and 2022. Estimated electricity capacity and generation timeseries between 1920 and 2022. London : His Majesty's Stationary Office, 2023.

15. Mc Garry, Toby. UK Electricity capacity and generation by fuel between 1920 and 2020. London : His Majesty's Stationary Office, 2023.

16. 2022 UK greenhouse gas emissions, provisional figures. 2022 UK greenhouse gas emissions, provisional figures. London : His Majesty's Stationary Office, 2023.

17. Department for Energy Security and Net Zero. Electricity since 1920. ondon : His Majesty's Stationary Office, 2023.

18. Ambler, Tim, Edwards, Peter and Clanmorris, Simon. UK Electricity Strategy: Capacity, Demand and Pricing. London : BEIS Select Committee, 2023.

19. Department for Transport. Road Traffic Statistics. A Department for Transport Web site. [Online] Department for Transport, 2023. [Cited: 1 December 2023.] https://roadtraffic.dft.gov.uk/summary#:~:text=Motor%20vehicle%20traffic%20on%20Great,to%20323.8%20billion%20vehicle%20miles.

IOR.org.uk

REFRIGERATION AIR CONDITIONING HEAT PUMPS

The knowledge hub for refrigeration, air conditioning and heat pumps

Heat Networks in Rural Areas

Dylan Betts

What you will learn

- The paper describes a pioneering rural village district heat network in Cambridgeshire using high temperature ammonia air to water and ground to water heat pumps to provide 2MW heat to 300 buildings including houses, school and two churches.
- The scheme is intended to provide space heating at 72°C water temperature to existing radiators through a Heat Interface Unit in each building and replaces mainly oil-fired boilers.
- The heat pumps are contained within a separate Energy Centre and include back-up electrode boilers and large thermal stores to provide resilience. Electrical Power is drawn from the Grid and a PV farm.
- The paper describes the issues experienced including community liaison, installation of the heat network distribution pipework and resolving unexpectedly high noise from the Energy Centre.

1. Introduction

The concept of rural heat network at the village of Swaffham Prior near Cambridge was first proposed in 2016 by the Swaffham Prior Community Land Trust (SPCLT). This was originally set up by members of the village to deliver housing for people living in Swaffham Prior who struggled to rent properties at typical market rental prices, or who needed to rent a different type of home.

Bouygues Energies & Services were appointed by CCC following the initial feasibility study to develop the technical feasibility of the project, before being appointed as Principal Designer to develop the detailed design and later as Principal Contractor. The Swaffham Prior Heat Network was conceived to represent an innovative method of renewable energy retrofit engineering and addressing not only the growing environmental and financial concerns surrounding oil and gas usage for domestic heating, but also the distinctive heating requirements of rural communities. The project developed around the concept of delivering hot water through a network of below ground pipes installed throughout the village to each property for both space heating and hot water. Heat pumps were selected as the preferred technology to supply the heat using both ground and air source technologies.

Two 750kW Ground Source Heat Pumps (GSHP), and one 500kW Air Source Heat Pump (ASHP) procured from GEA Refrigeration UK Ltd and operating on Ammonia (R717) refrigerant have subsequently been installed in a new purpose-built Energy Centre built upon the site of a previous agricultural building.

The GSHPs are served by an array of 103 boreholes, located in the adjacent field to the Energy Centre. Four $50m^3$ thermal stores sit outside the building, these enable the decoupling of energy production from time of use, enabling the heat pumps to take advantage of time of use tariffs of the availability of cheap electricity. As a further innovation, Bouygues proposed to provide an electrical connection from a nearby proposed Solar PV farm to the Energy Centre, provide renewable electricity for the heat pumps. The network will have the capacity to supply 2MW heat generating capacity of heat from the heat pumps at around 72°C to around 300 properties in the village. This value had been determined using previous years' energy consumption from a sample of the properties in the village. The majority of the properties are domestic dwellings, but the village also includes two churches, a school and a pub. Despite the use of previous years' consumption data from various properties in the village, baseline data could not be accurately measured due to the lack of metered data, the variability in use across properties and a distinct lack of reliable assessment techniques for rural buildings. Additionally, pro-active decisions had to be made regarding the best method of predicting future energy usage of the village over the 60 year life-span of the project. However, fluctuations which will arise dependent on demographic, age, and familial nature of future residents provides a huge variability in predictive energy consumption for the future. As such, the thermal buffer tanks serve as additional heat generation capacity for the network, providing a further 2MW of peak heat as required by the village. The design was put together following appropriate design philosophies and guidelines set of in CIBSE CP1, however this didn't include for rural challenges, and thus, suitable alterations had to be made.

It is necessary to contextualize the significance of this endeavour within the broader environmental landscape. Rural households contribute 25% more CO_2 emissions from heating fuels than their urban counterparts, amounting to an additional 1.5 tCO_2 per household per annum [1]. This stark contrast accentuates the pressing need for innovative solutions, such as the Swaffham Prior Renewable Heat Network, to mitigate the environmental impact of rural communities.

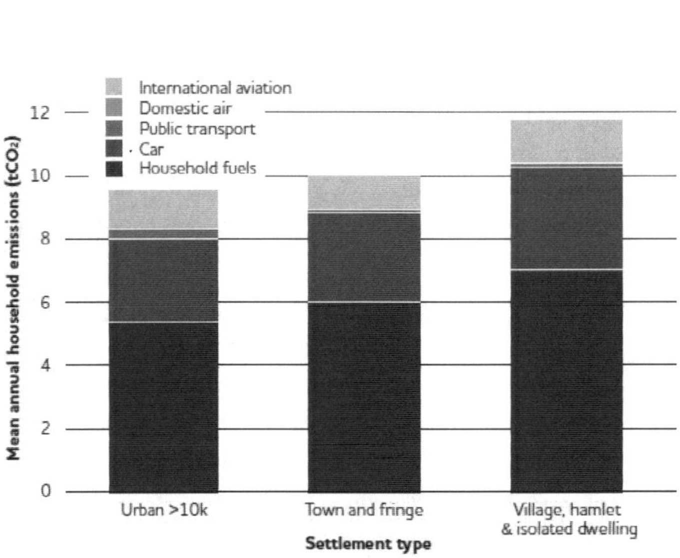

Figure 1: Mean Household Emissions by Settlement Type [1]

This paper aims to delve into the utilisation of Heat Pumps within the Swaffham Prior renewable heat network and aims to shed some light into the complexities and challenges associated with attempting a large-scale retrofit of heating systems in a village of approximately 300 properties. By examining the lessons gleaned from this groundbreaking initiative, the paper will explore the potential of referencing Swaffham Prior as a compelling case study for analogous village retrofit programs throughout the UK. Through a comprehensive analysis of the technological, logistical, and social considerations underpinning the project, this paper seeks to offer valuable insights into the viability and implications of integrating further sustainable heat networks in rural areas throughout the UK, thereby contributing to the necessary advancement of sustainable heating practices nationwide as per both the "Building for 2050" initiative [2] and the UK Government national carbon emission reduction targets of 2030 and 2035.

2. Project Background

The Heating Swaffham Prior project was kick-started by Swaffham Prior Community Land Trust to give residents the opportunity to cost-effectively replace their oil-fired heating systems with a low-carbon alternative. In 2018, the trust commissioned a feasibility study, funded through a £20,000 grant under the Rural Community Energy Fund, into alternative heating options from sustainability charity Bioregional, which identified the potential for a village-wide heat network. A range of heat sources was considered – including burning biomass from local straw – but the key to unlocking the project was a land parcel on the edge of the village owned by Cambridgeshire County Council (CCC). This opened up the possibility of generating heat using a GSHP system.

The Phase 2 Rural Community Energy Fund grant was only available to district and parish councils, as a result, the Community Land Trust presented the heat network concept to Cambridgeshire County Council with a view to persuading the Council to apply on their behalf for a grant towards project design and development. Following this, Cambridgeshire County Council approached Bouygues E&S through the existing REFIT 3 Energy Performance Framework to develop the project further.

The project was initially intended to utilise Ground Source Heat Pumps only, with approximately 267 boreholes, however the project scope was altered as a result of a sharp decrease in the previously preferential Non-Domestic Renewable Heat Incentive (NDRHI) tariff rates alongside an significant increase in the cost of borehole installation and commissioning. The addition of the single 500kW Air Source Heat Pump for use in the summer months was incorporated to give time for the residual heat underground to recover, so that, as the 20+ year programme continues on, the thermal low point of the ground from which the boreholes extract heat will be considerably higher year-on-year relative to a system which only operated Ground Source Heat Pumps throughout the year.

2.1 Community Engagement through the Swaffham Prior Community Land Trust

Community engagement has been vital throughout the project development process, the construction phase and ongoing operation of the heat network, and as such, a lot of time and effort was invested in engagement work to boost the confidence and trust of the community who were being asked to commit to switching to a new type of heating provision which the majority had little knowledge of. As part of this effort, regular stakeholder meetings were held with representatives from CCC, Bouygues E&S and the SPCLT in an attempt to respond to concerns and answer any questions from the village residents. Regular surveys were conducted to gauge resident opinion of relating to the quality of customer service, specifically regarding knowledge transfer on how the heat network was intended to operate, the technologies that would be employed, receptiveness to resident complaints and queries, and full transparency as to what negative effects could be experienced as a result of the installation of the heat network. To further improve stakeholder relationships, a stakeholder engagement consultant was engaged through the Community Land Trust to further educate the residents as to the benefits of the project.

2.2 Community Reaction and Further Engagement

As of January 2024, there are approximately 70 properties connected to the network and "on-heat" with an additional block of properties in the planning phase. Of those connected, the vast majority have been very satisfied with the level of heat provided and the ease of use of the newly installed Heat Interface Unit (HIU) which essentially takes the place of their previous boiler. This serves as the 'interface' between the heat network and the tertiary heating distribution system in the property. In addition, due to the majority of the properties in Swaffham Prior originally operating on an oil fed system, the absence of an unsightly oil tank is a welcome sight in their back gardens. Community support post-installation is equally important to the success of the project, and to that end Equans E&S operate a 24-hour hotline for Swaffham Prior residents in the event of a loss of heat, similarly, the operation of the energy centre is monitored for 24 hours a day.

There have been some unforeseen issues regarding the construction of the network which the community has fed back to Equans E&S, CCC, and the SPCLT. These concerns are being addressed by the team dedicated to the operation and maintenance of the network, located on-site at the Energy Centre. In summary, these issues are:

1. The visual impact where grass verges have been poorly re-filled following groundworks.
2. The visual intrusion of the Energy Centre for homes in the village (although this is approximately 0.5km from the village centre).
3. The original footpath which passed through the borehole field, has only recently become available for use.
4. Difficulties caused as a result of necessary easements through unconnected properties.

The residents of connected properties have also experienced some issues concerning recent works post-installation of the HIU and connection onto the heat network. These have often reflected a limited knowledge of the new system from the residents and have affected water circulation related works, for example one resident reported flow issues following a bathroom re-fit. This, as is the case with many resident queries can be addressed with a higher level of focus on resident education for the future.

Furthermore, the case of responsibility delineation has been prevalent in the 18 months since the network was turned on and the first connections made; an example of ongoing discussions between the project partners concerns the replacement of 'failed' heat emitters within homes subsequent to connection to the network. Equans E&S are of the opinion that if no modifications have been made to the existing physical emitters and pipework during the installation works, it is the duty of the homeowner and/or CCC to resolve, unless clear evidence can be provided that the installation works that cause the damage that has been reported. These are often complex and difficult conversations where it is of paramount importance for all parties to work together to reach an amicable solution. Clear delineation of responsibility, before any works were started, could potentially be improved to prevent such issues from happening.

3. Energy Centre

A purpose-built Energy Centre has been constructed to locate the 2x 750kW GSHPs, 1x 500kW ASHP, 3x 500kW auxiliary electric boilers, dedicated ventilation equipment (due to the use of ammonia as a refrigerant), and all other heating ancillary equipment. Due to their size, the thermal stores are sited externally with the dry air coolers which extract heat from the ambient air for the ASHP. The Energy Centre is situated on the eastern side of the village and was designed and procured by CCC independently although works were sequenced and coordinated with the design packages managed by Bouygues E&S in our role as PC/PD.

Figure 2: Energy Centre (left) and Thermal Stores (right)

The Energy Centre was built with a 12m^2 roller door installed on the northwestern face of the building, a factor which has proved to have a negative impact upon noise levels, and contains internal access to the Northeast of the building. Within the energy centre there are three separate fit-for-purpose control rooms alongside the bulk of the Energy Centre where the heat pumps are located, along with 3x 500kW Electrode Boilers which were included in the overall heating design to provide additional resilience. These can provide the thermal stores with heat if the heat pumps are offline. Although these are clearly inferior in terms of overall efficiency relative to the GSHPs and ASHP, the opportunity to secure a significantly sized electrical connection (via the Cambridgeshire Private Electricity Network) made this a viable and practical option. Fig.2 shows an external view of the Energy Centre where the 4x 50m^3 (200,000 litres total volume) thermal stores are located. These are used to buffer thermal heat at 72/75°C from the heat pumps / electric boilers to meet demand from the network.

An agricultural building used to be located on the site previously and as such, planning permission did not have to be sought in order to complete the build. We acknowledge that there have been a number of issues to address regarding the Energy Centre since system operations have begun; perhaps the most pertinent of which has been the noise level at the border of the single adjacent property. The noise levels tend to be more noticeable as a result of operation of the Ground Source Heat Pumps at night. Noise levels were planned to be approximately 25dB initially, with this rising to 45dB during the project planning phase. Actual noise levels can exceed this, and a review of acoustic absorption methods was considered.

As a result, acoustic jackets have been fitted over the compressors to reduce vibrational noise emissions. However, insufficient noise attenuation measures from the building overall, were implemented during the construction phase causing small gaps to be left (for example around the roller shutter door) and a measured noise level at the neighbouring border in the order of 50dB. In addressing these higher-than-expected noise levels relative to the original scheme design, the heat pumps have all been given increased layers of sound absorbent protection and all aforementioned air leaks in the building fabric have been dealt with. Acoustic membrane has been applied to all external doors, with increased layers on any louvre doors; cumulatively these preventative measures are expected to reduce the decibel level at the border of the neighbouring property by 7-9dB and will be an important achievement in ensuring addressing the concerns of all those potentially impacted by the heat network's implementation. Although this was identified as a project risk early on during the development phase, noise levels have proved to be in excess of expectations and serves as an important issue for consideration in any comparable future projects.

4 Heat Pumps and Refrigerants

Ammonia heat pumps have been utilised for the scheme as primary heat sources. These needed to be capable of delivering 75°C water flow temperatures at a delta T of 15°C. The heat pumps featured a twin stage compressor circuit comprising of a screw compressor with a reciprocating compressor for high-end temperatures. The 500kW ASHP is connected to a dry air evaporator cooler (located on a plinth directly outside the building) to absorb ambient thermal energy and transfer via the refrigeration circuits to the heat pumps internally.

The possibility of using R717 as a natural refrigerant alternative (with a GWP of 0), came to light through an extensive market evaluation of available commercial heat pumps during the early project design phase. It became clear that in order to get to the necessary temperatures that would align to a conventional fossil-fuelled boiler system as found in the majority of homes, we would need to utilise a slightly more 'unconventional' refrigerant option relative to HFC refrigerants that are more commonly found in commercial applications. After initially looking at HFO refrigerants, these were unable to reach the expected high temperatures at a reasonable compression, which would result in decreased COPs as the required temperatures are met. There are also wider environmental concerns with the use of HFO's. As a result, natural refrigerants with ultra-low GWP's were considered, with CO_2, Propane, Isobutane, and Ammonia as the main options.

CO_2 was deemed unsuitable due to the requirement of a very low return temperature on the heating side in order to maintain the stability of the vapour compression cycle, the refrigerant would not be able to deal with the narrow delta Ts attributed to the Swaffham Prior Project. The use of CO_2 would have required a +40°C delta T which could not have been retrofitted into the properties on the village without severely impacting the secondary heating systems.

Ammonia was chosen over Propane and Isobutane due to its commercial availability within the heat pump market at the time, however all were suitable refrigerants. The drawbacks of these are that they are hazardous and potentially explosive. Ammonia produces additional risk through its toxic and corrosive nature and the risks associated with inhalation. Due to the safety issues with the ammonia refrigerant, extra precautions were taken in the design and construction of the energy centre in which the heat pumps were housed to ensure full compliance with DSEAR regulation practices. These included ventilation and leak-detection systems as well as various other safety measures to ensure that – in the event of a leak – there is a process set up to keep any personnel safe and avoid an explosion.

The use of both Ground and Air Source heat pumps allows the system to operate using the heat source which provides the best co-efficient of performance at the time as monitored by the BMS in the Energy Centre.

4.1. Borehole Field

The arrangement of the borehole field was altered significantly throughout the project. Initially Bouygues commissioned BA Hydro to conduct hydrothermal geological studies of the site to enable understanding and prediction of heat generation potential of different borehole arrangements. Also undertaken was a thermal resistivity test to determine the volume of heat able to be absorbed from each borehole, this was achieved through the installation of a test borehole with thermal response tests carried out over several weeks to inform the borehole design through improved understanding of the geology. This led to the inclusion of an air source heat pump to relieve the boreholes of their duties over the summer months and replenish and regenerate heat for the following winter.

Subsequently, this led to the ability to have the boreholes dug in a far sparser pattern than originally intended, which in turn also improved efficiency of the boreholes as there was far less interaction between each neighbouring set of boreholes. This allowed a further reduction in the total number installed since each borehole was able to be exploited to its full potential. This iterative process between the M&E designers, energy specialists and thermo-geologists allowed an optimum combination to be arrived at between the use of the air and ground source systems. As a result of our modelling, including the optimisation of design and control philosophies and increasing thermal storage capacities the planned number of boreholes was reduced from 256 to 103.

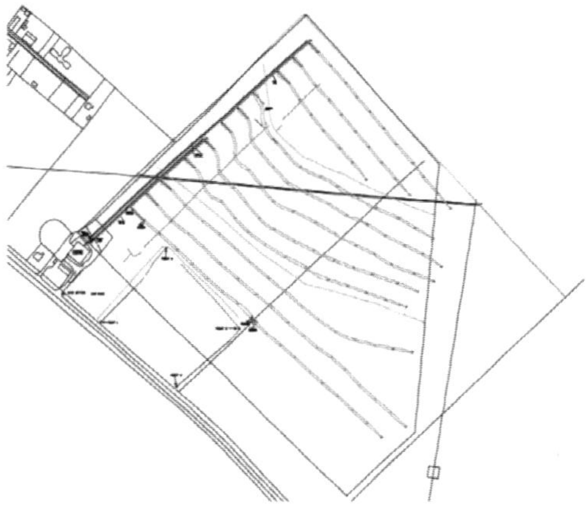

Figure 3: Borehole Field Layout

The boreholes are connected to the energy centre and heat pump exchanger via a main branch of flow and return pipework which is fed by individual strings (rows) of pipework stemming from each row of boreholes – as seen above (Fig.3). The control of each junction between the main branch of pipework and the individual strings house a manifold which has the capability to manage flow rates and pressure through the system.

5. Heat Network and North Angle Solar Farm

5.1 North Angle Solar Farm (NASF) / Cambridgeshire Private Electricity Network (CPEN)

The increase in electricity prices in recent years and the design and construction cost for the network had a substantial impact on the feasibility of the project and placed a strain on the economics case, given that the price of the commodity being relied upon to deliver this work was increasing to north of 60p/kWh. An additional strain to the project consisted of UKPN costs to upgrade the connection to the national grid from the Energy Centre. Rural villages do not typically have large existing connections to the grid – and not enough to draw an additional 2,500 kVa that was required to the new Energy Centre. As a result, Bouygues E&S proposed connecting to the North Angle Solar Farm - which at the time was an independent Bouygues E&S project. A cost-plan and business case were developed for the installation of a private wire connection from the North Angle Solar Farm to the Energy Centre to supply the electricity demands of heat sources. Using electricity with 'zero carbon' emissions, when compared with the carbon emissions contributing to grid electricity, is extremely beneficial and it was agreed to include this in the overall project proposal.

CPEN is an 8km long stretch of 33kV private wire with a central substation directly connecting NASF with a new transformer just outside the Energy Centre. Anything downstream of the central substation is private, linking directly to the energy centre but allowing for surplus export to the network when there is little demand, and shortfall import from the network in the event of increased demand during periods of low solar absorption.

5.2 Heat Network Layout

Bouygues E&S contracted with Pinnacle Power Ltd for design and build of the heat network. Designing and installing a new heat network is an unprecedented challenge when retrofitting an ancient parish which contains houses built as far back as the 1700's [3]. The DHN was designed with a 70°C/55°C degrees Celsius flow/return design operating temperature, to ensure efficient operation of the heat pumps and thermal stores, however design of a flow/return system with a delta T of 15°C presented a unique challenge with regards to the arrangement of the older residential properties. Typically installed heating systems in the village operated at 10°C delta T, which resulted in the return of water through the network at 5°C more than designed. To mitigate this, Bouygues E&S considered replacement of the pipework distribution system in the properties in order to satisfy the 15°C delta, however the high costs and potential disruption involved made this an unsuitable option. The network itself – as shown in Figs.4 and 5 comprises of an intricately laid series of flow and return pipes throughout the roadways of the parish. The network design allowed for future connections in all homes in the village as well as the small number of existing non-domestic properties. The wider project team including CCC felt this was important since it would provide the opportunity to connect to the network at a later date. At each property line a t-connection was installed for later use as a branch off to the property line.

The network itself – as shown in Figs.4 and 5 comprises of an intricately laid series of flow and return pipes throughout the roadways of the parish. The network design allowed for future connections in all homes in the village as well as the small number of existing non-domestic properties. The wider project team including CCC felt this was important since it would provide the opportunity to connect to the network at a later date. At each property line a t-connection was installed for later use as a branch off to the property line.

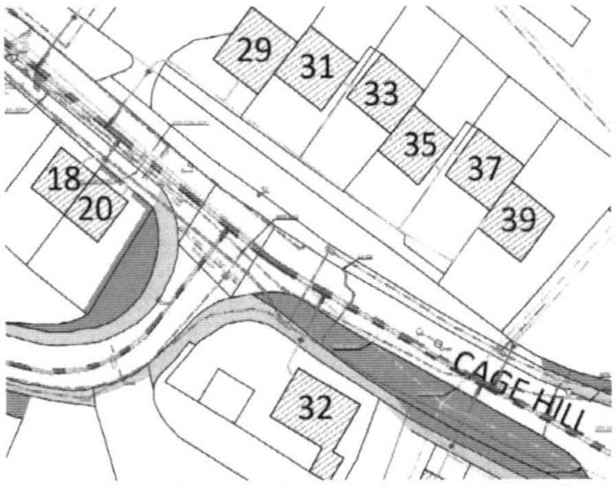

Figure 4: Heat Network - T-connections at property line

Figure 5: Entire Heat Network / Energy Centre (shown in purple)

5.2.1. Network Installation & Traffic Management

Pinnacle were sub-contracted to install the network throughout the village. Traffic management proved to be a particular area of concern throughout the installation process. The rural nature of the village meant that unforeseen challenges arose with respect to planning and reception of right to work permits on the roadway. Where possible permits were acquired in advance for the initial installation, however for end-point customer connections this wasn't possible as the installations were not delivered to every property in the village sequentially, which would have allowed for a more organised approach for installation and traffic management. Instead, permits had to be obtained after residents had signed an "Expression of Interest Form" and several surveys had been carried out on the property to determine pipework routes from the roadway into the home. All traffic management attributed to end-customer connection had to abide by several notable caveats exclusive to projects in rural areas, such as the High Street – main road through the village – being part of an access route for HGV farm deliveries which require additional space on the roadway than do ordinary vehicles.

5.3. End-Point Customer Connection and Internal Installations

The process by which a resident of Swaffham Prior is able to join the heat network and have the pipework routes designed and installed on their property is designed to be as streamlined and un-intrusive as possible for the homeowner. Following a property owner signing an expression of interest form with Cambridgeshire County Council, the process involves Equans E&S conducting an internal and external survey on the property. This includes aspects such as:

1. Determining make, model and age of the existing boiler
2. Determining kW rating, flow and inspection history of the existing boiler
3. Carrying out inspection surveys on the existing boiler controls
4. Performing a visual inspection of building electrical systems to determine the need for a new distribution board on the property as a result of the increased electrical load.

Following this process, a surveyor locates the suitable area for the network pipes to enter the property, this is often where the existing boiler is already located, in order to facilitate a straight swap with the heat interface unit. Then a survey is undertaken to provide pipework routes from the T-piece in the roadway to the home with measurements and costs produced and sent to the Council for review.

Following the review, the property is factored into the construction programme. Works are then carried out later; over the course of 4-5 days the pipework is laid with all mechanical and electrical connections in place and the new HIU installed with a commissioned data logger installed which will send consumption information back to the energy centre to better assess and improve output efficiencies.

6. Swaffham Prior Heat Network as a Case Study for Retro-fit Heat Network

The Swaffham Prior HN is a case-study for retrofitting and rejuvenating ageing villages for them to become more accessible both to newer-generation renters and growing families whose preference is to not reside in a home fuelled by oil/gas. By instead fuelling the homes with zero carbon alternatives such as a mixed Ground and Air Source Heat Pump Network.

An important aspect of the success of the project came through weekly design team meetings involving designers and engineers from a plethora of companies and disciplines. Listed below are the main commercial partners involved in the project:

- Sharman Grimwade (M&E Designers)
- Pinnacle (DHN)
- BAS Hydro (Boreholes)
- GEA (Heat Pumps)
- Bouygues E&S

Whilst the project has largely been a success, the challenges outlined in this paper will serve as a forethought to those who wish to replicate the heat network regarding the logistical and technical issues both foreseen and unforeseen that have had an impact on the development of the SPHN.

About the author

Dylan Betts, BEng (Honours) is a graduate in Energy Engineering from De Montfort University. Dylan has been involved in several large-scale decarbonisation projects over the last year including the implementation and operation of the Swaffham Prior Heat Network project, which is the first of its kind in the United Kingdom, providing the capacity to have up to 300 homes decarbonised with low carbon heating systems

References

1. V. W. J. T. a. T. B. Ian Preston, "Distribution of Carbon Emission in the UK: Implications for Domestic Energy Policy," University of Oxford, Oxford, 2013.
2. E. &. I. S. Department for Energy Security and Net Zero and Department for Business, "Building for 2050," London, 2018.
3. "Swaffham Prior," British History Online, [Online]. Available: https://www.british-history.ac.uk/vch/cambs/vol10/pp273-278#anchorn9. [Accessed 22 01 2024].

Presented before the IOR on 7 March 2024

Q & A - Challenges and opportunities for heat pumps in the domestic housing market

Dermot Cotter – From your presentations it seems to be complicated to do anything with heat pumps. When you look at heat pumps and the cost are you finding that a single domestic heat pump or a heat network costs less in the long run?

Chris Parker – I can't answer that but did try to investigate it during my research. We know roughly the cost of an air source heat pump and we know what the government plans are for heat pumps in the future but a value that I have been struggling to find is how much does it cost to connect a house to a heating network? As a network grows out, the further away you start to connect properties, they become more expensive for pipework etc so at the moment it's hard to find a figure to say it will cost x amount of £ to connect a property to the network.

Dylan Betts – What is important to realise from a network point of view is that the infrastructure is all installed at once, whereas if 300 households were to install separate heat pumps it would require individual UK PN connections and more extensive individual pieces of construction in the public roads. I think that's a benefit of a public heat network as opposed to air source heat pumps installed individually. I'm not entirely certain of costs and I wouldn't want to comment and be wrong. There's definitely benefits to both, In rural areas individual heat pumps would possibly not be the most efficient option but in cities that would perhaps be a better option.

Dermot Cotter – Chris, you brought up older houses and lack of insulation. Within your findings, do you believe that it is the people who can least afford to update their houses?

Chris Parker – I believe it is. I don't have anything to back that up because that is the bit of missing data that I referred to in the presentation. We know the population, we know the income for the population and we know the overall housing but we don't know where those lines cross. It's a very important piece of information to know how many low income houses and where they are in the property market but from our own experience of being on the streets and exploring areas we can usually tell that the more deprived areas typically have older houses, which would usually indicate a lower income threshold.

Dermot Cotter – When you drive around the country you also see older houses in rural settings. Dylan, did you see a mixture of housing in rural settings within your project?

Dylan Betts – The difference between the houses in the villages is something that has to be considered in the design. There are a lot of houses which are towards the earlier end of the time period that Chris spoke about and some houses that have been recently refurbished or even rebuilt and to that end there is no 'one size fits all' heating model for the network. It's more based on the end customer's usage than the ability of the heat network to provide more heat to the house depending on it's age. I think it's still very much left to the homeowner.

Dermot Cotter – The temperature of a heat network is 75 degrees and as a principle for every 1 Kelvin higher that you heat something it is 3% more energy use so if it was possible to reduce it down to 70 degrees then it would be much less energy use for the heat pump which is where you want to be going in the long run. How would you get a village like that to slowly reduce that temperature? If you have a house that requires 75 degrees and everybody else only needs 55 degrees then that house is costing everyone extra.

Dylan Betts – I think if it came to that then we would consider refitting the pipework and heating system in that house. For a whole village of 300 houses and numerous commercial buildings that wouldn't be feasible but for 1 – 10 properties that would be an effective business case to do that.

Dermot Cotter – When it comes to the cost of heat pumps, what would you say is the average cost of putting in a heat pump in a house?

Chris Parker – The average cost, at the moment, is usually quoted to be around £14,000. That is the capital cost of the heat pump unit and the cost of getting experienced, trained professionals to fit it and strip out the old gas boiler. That would be assuming no replacement of pipework or radiators. As soon as you start looking at whether radiators and pipework are suitable that cost may go up even more. There's a lot of reference to microbore pipework being unsuitable for the flow rates needed for heat pumps and as soon as that needs to be replaced that becomes a very disruptive installation, going into walls and pulling up floorboards to replace pipes. How much the cost increases will depend on the size of the house and how many radiators there are. There are government grants to help with the cost up to £7,500 but as time goes on we should see that overall cost come down as more professionals are trained and if the government is able to achieve their price reduction targets then a heat pump capital cost should come down to £5,000 by 2028 and on parity with boilers by 2030 then we should hopefully see much less expense for future installation. One worry though is that to make costs on a par with gas boilers you could increase the price of gas boilers so you don't actually ease the cost burden, you just make it even more difficult for everyone. There was a proposal to do that this year but I believe that was dropped so we shouldn't be seeing that effect.

Dermot Cotter – Dylan, when it comes to your energy centre. Do you have people there all the time or is it manned just a few hours a day?

Dylan Betts – We have staff there between 8am and 6pm but we also operate a 24 hour helpline for call outs so any customer connected to the network can phone that helpline at any time and a technician will be sent out to them to fix the problem if it can't be done over the phone.

Dermot Cotter – There are approx. 70 houses on the network. I imagine if there were 70 boilers in houses then they would not have this kind of support doing maintenance. Therefore, is it financially cheaper to buy heat from the network or to have oil powered boilers.

Dylan Betts – I'm not entirely sure whether for customers it provides significant savings over oil. Because you are paying direct for the oil that you are using you are recovering the cost of the project through your monthly bills. I believe that as technology advances and more heat pumps are installed that the price will come down and that it will become considerably cheaper.

Dermot Cotter – What I am hearing is that this is all very complicated. Doing a domestic house is complicated and doing a heat network is also complicated. In a way putting a rural heat network in seems slightly easier than converting domestic houses and all the problems associated with replacing pipes etc because you can use existing pipework. Am I correct?

Chris Parker – It depends on what type of pipes you have. The cost of disruption depends on the size of the house. The bigger the house, the bigger the disruption.

Dylan Betts – For the Swaffham Prior heat network there were very minimal changes to the internals of the homes. It was considered, but judged against the business case, it was deemed unsuitable. That's not to say that it wouldn't be necessary for any other heat networks in rural villages.

Dermot Cotter – For Swaffham Prior you used a combination of ground source and air source heat pumps. Was there any consideration given to putting in a combination heat pump that could work from the borehole in the ground or from the air. Why did you put in two separate systems rather than one combi system?

Dylan Betts – Originally the idea was to just have a ground source array but what was realised was that over the course of the payback period of the project the heat harnessed from below ground would have dropped considerably over the course of the 50/60 year time period because the ground would not have been given enough time to regenerate it's heat. The use of the air source heat pump allows for a complete switch off of the ground source pump over the summer months and, as such, the bore holes have 4 to 6 months to regenerate heat and provide enough for the winter.

Dermot Cotter – My question was more why didn't you put in a hybrid heat pump that could take heat from the ground or the air rather than using two separate heat pumps.

Dylan Betts – I don't know.

Dermot Cotter – It would be quite a simple thing to do. Do you have any closing remarks Chris?

Chris Parker – We need to openly discuss these challenges going forward, not just within our own industry but beyond. The amazing thing that the IOR can do as an association body with experts from the industry and what we can also do as individuals is inform the people who make the decisions who don't necessarily know the technicalities. We can't expect MPs to be experts. We need to keep talking.

Dermot Cotter – And you Dylan?

Dylan Betts – An echo of Chris' initial point of lobbying and working together. One of the main points of my conclusion was the effectiveness of the cohesive team that was built between all those involved in Swaffham Prior and for future advancements in domestic heat pump installation or renewable energy needs to come from creative teamwork and connections between businesses, councils and community trusts.

IOR.org.uk
REFRIGERATION AIR CONDITIONING HEAT PUMPS

The knowledge hub for refrigeration, air conditioning and heat pumps

Emissions Assessment of Refrigerated Food Road Transport in the UK

Dr Rasaq Lamidi, Dr Xinfang Wang, Dr Hameed B. Mahood, Prof Toby Peters, Dr Leyla Sayin, Dr Ana C. Marques

Why you should attend

- To gain insights into estimated energy use and emissions from road transport refrigerated units in the UK
- To have a better understanding of vehicle numbers and composition
- To understand trends in consumer shopping and how they affect the demand for transport refrigerated units

Abstract

Above 60% of food in the UK is dependent on the cold chain and an estimated 2 - 4% of the UK's total greenhouse gas (GHG) emissions is associated with food refrigeration [1]. Refrigerated vehicles are essential parts of the modern cold chain distribution system, particularly in linking farmers and consumers while extending perishable food's shelf life and helping meet the required regulatory temperature constraints. Transport Refrigeration Units (TRUs) are presently primarily powered by diesel fuel. The UK government is committed to Net Zero emissions by 2050 from all sectors, including transportation, hence decarbonisation of this sector is paramount.

The present paper aims to estimate the energy demand and emissions of the UK temperature-controlled transport using a bottom-up approach. Only energy and emissions associated with the TRUs of road refrigerated transport are considered, while traction-related fuel is excluded. Depending on their sizes, refrigerated vehicles were categorised into cargo (motorcycles and tricycles), vans (<3.5 tonnes), trucks (>3.5 ≤ 18 tonnes) and trailers (>18 tonnes).

The study estimates the number of temperature-controlled vehicles in the UK to be around 158,934 ± 46,326. This includes 4,391 - 8,004 E-cargo bikes; 44,877 - 81,802 vans; 21,561 – 39,301 trucks and 41,778 - 76,152 trailers. Depending on the number of vehicles, the total annual energy and emissions associated with transport refrigeration units varied between 5.13 - 9.36 TWh and 1.44 - 2.62 $MtCO_2e$ respectively. Scope I emissions vary between 1.44 to 2.62 $MtCO_2e$ (0.20 - 0.36 $MtCO_2e$ from refrigerant usage and 1.23 - 2.26 $MtCO_2e$ from fuel use). The Scope II emissions which is the emissions associated with electricity consumption also varied between 446.2 and 813.4 tCO_2e.

1. Introduction

By 2030, the expected population growth could lead to a global increase in food production and energy demand by about 50% and 45%, respectively [2]. Meanwhile, climate change has imposed additional challenges to the already over-exploited finite resources' capacity to meet the projected growths. To avoid the worst outcomes of climate change, the world must limit global temperature rise to 1.5°C and attain Net Zero emissions by mid-century. The food and drink sector has a significant role to play since food systems emit about 15.8 $GtCO_2e$, equating to 30% of the world's GHG emissions. In particular, the global food miles account for about 3.0 $GtCO_2e$ of the total food systems emissions [3].

Therefore, governments worldwide, including the UK, have prioritised food security and resilient food systems. Consequently, the UK government has envisioned sustainable and secure food systems in 2030 in response to demographic and climate changes. The strategy mainly centres on reducing the resources used in the food system, reducing GHG emissions, and minimising food waste while ensuring the food system's competitiveness [4].

Meanwhile, "the farm to fork" chain in the UK is worth £120 billion, with the food and drinks manufacturing sector being the largest UK manufacturer, contributing £29 billion to the economy yearly and employing over 440,000 people. The UK's food system relies on imports; about 84% of fruits and 47% of vegetables were imported in 2019 [5]. Thus, making the UK food system dependent on transportation.

The literature suggests there is an increasing preference for frozen food consumption, primarily driven by longer shelf-life, lower price and preference for lower shopping frequency [6]. [6] For instance, within the UK, fresh products sales value dropped by 8.3% while frozen food sales value increase by +9.5% between 2022 and 2023 [7] and with right information on waste reduction and nutrition values about 42% of consumers are willing to buy more frozen foods [8]. According to the UK Cold Chain Federation (CCF), there is approximately 51% increased demand for products that pass-through temperature-controlled supply chain with a corresponding 62% increase in demand for environmental sustainability [9]. Therefore, there is increasing demand for the temperature-controlled transportation system. Refrigerated transport has the twin burden of emissions from traction and TRUs. Emissions in the transport refrigeration sector include direct emissions of refrigerants (fugitives) with high global warming potentials and emissions for powering the TRU, usually vehicle engine, separate diesel driven TRU or from electricity used to power TRU [10].

Understanding how many temperature-controlled vehicles are on the road and their contribution to GHG emissions is essential for planning appropriate and effective mitigation measures. It will also help monitor emissions reductions and substantiate policy measures once a good baseline has been established. Unfortunately, there is no accurate data on the number of refrigerated vehicles on UK roads. For instance, while CENEX [11] puts TRUs on the UK road at 100,000, Foster et al. [12] suggest 70,000 TRUs on the UK roads. Therefore, this paper aims to estimate the current fleet size and compositions of the temperature-controlled vehicles on UK roads. Using a bottom-up approach, including a survey of key stakeholders, the result is then used to estimate energy and emissions associated with the temperature-controlled road transport of the UK baseline year 2023.

2. Methodology

2.1 Vehicle size and composition

The analysis of fleet size and compositions of temperature-controlled supply chain in the UK and subsequent assessment of associated energy and emissions are done following the approach detailed in Figure 1. The methodology included a desk-based study of existing literature to understand and analyse fleet size, compositions, and stakeholders involved in temperature-controlled transportation within the UK.

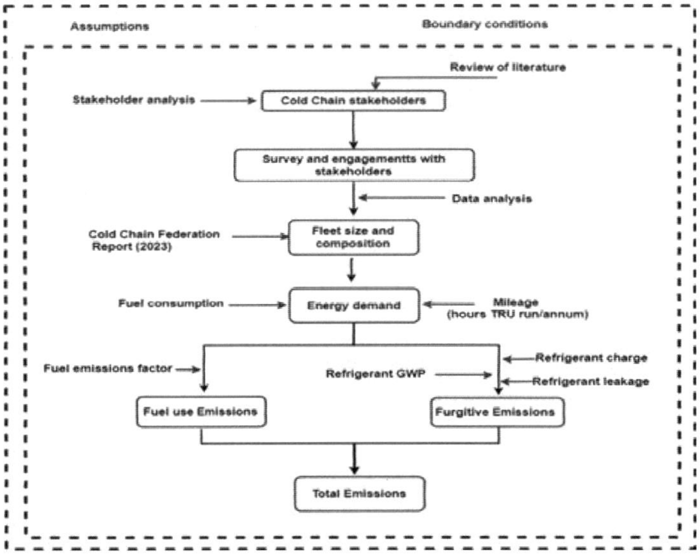

Figure 1: Approach to energy and emissions analysis of TRUs in the UK.

This was then followed by stakeholders' engagement and survey. The survey was conducted between 29th June and 1st September 2023, and the survey questionnaire was sent to 139 businesses across identified stakeholders. Businesses with direct control of refrigerated vehicles are broadly categorised into four groups: Transport operators (supermarkets and food distribution), commercial bodybuilders, leasing and rentals, and TRU Original Equipment manufacturers (OEM) and Dealers. Of the 139 businesses, 52 are transport operators, 23 are commercial bodybuilders, 44 are leasing and rental companies, while 20 are TRU-OEM businesses.

The survey was conducted through SurveyMonkey, and tailored questions were sent to various categories of stakeholders based on the nature of their business within the food cold chain system. For instance, questions sent to the bodybuilders differ from those sent to the TRUs equipment manufacturers. However, the questions centred on TRU numbers, composition (vans, trucks or trailers), fuel consumption, mileage, number of compartments, vehicle utilisation capacity, vehicle usage, renewable energy adoption, etc.

Table 1: Assumptions and boundary conditions for modelling

Boundary condition	Only the emissions related to the TRU were considered, while emissions associated with traction were excluded.
Other boundary conditions	• TRUs are predominantly powered with diesel, which was considered as the main fuel for the modelling. • The literature indicates that R404 is the predominantly used refrigerant. However, our survey suggests that R134a, R452A, R410A, R1270 and R744 also exist in small proportions. Hence, a common GWP is calculated as a function of the identified fractions. • Refrigerant leakages for various sizes of temperature-controlled vehicles are as indicated on HFC Outlook Model [13] and Francis et al. [14 • Electricity emission factor = 0.207 kgCO2e / kWh • The conversion factor for diesel is 2.63 kg CO2e per litre of diesel. • Vans are direct drive and only use a portion of fuel for the refrigeration system [15]. Previous empirical studies indicate this in the range of 14.5% to 25% of traction fuel [10,16]. Hence, 14.5% is assumed for vans.

3. Results and Discussion

3.1 UK food cold-chain and stakeholders

The food cold chain in the UK is complex and interrelated, involving production, packaging, processing, warehousing, wholesaling, and retailing with temperature-controlled transportation linking all stages.
It also involves many active stakeholders (Figure 2), including vehicle manufacturers, bodybuilders, TRU original equipment manufacturers, retailers, food transport and distribution companies (Third Party logistics companies), food services enterprises, regulators, trade associations, etc. Typically, a temperature-controlled business approach bodybuilders with a vehicle needs and business requirements, e.g. size, temperature requirements, and compartmentalisation. The bodybuilder then advises on suitable chassis and TRU and then couples all the elements together and ensures the refrigeration box meets both business requirements and regulatory constraints, e.g., the choice of refrigerant and K-value box insulation system.

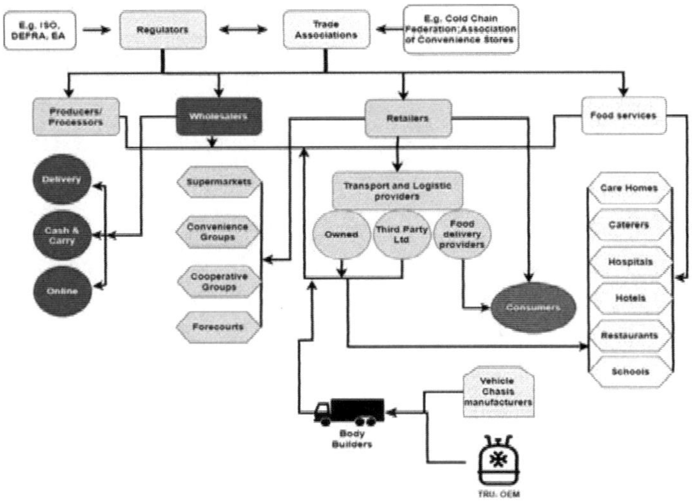

Figure 2: Temperature-controlled transport stakeholders.

3.2 Fleet size and compositions.

There were 26 responses from the survey, representing a response rate of 19% from the identified 139 stakeholders. Based on the fleet size, respondents control 36,181 temperature-controlled vehicles (TCV). This includes 863 TCV from bodybuilders, 7,952 from Lease and Rental service providers, 6,175 from TRU-OEM and 21,191 TCV from food distributors and retailers.

Our analysis of the fleet composition (Figure 3) from the survey indicated an estimated 49%, 10%, 37% and 3.9% for vans (LCV less than 3.5 tonnes gross weight), trucks (HGV greater than 3.5 tonnes but less than 18 tonnes gross weight), trailers (HGV greater than 18 tonnes gross weight), and e-cargo bike respectively. The Cold Chain Federation report of 2023 [9] suggested their 278 members, predominantly in the food storage and distribution sector, operate 62,730 TCV in the UK. This includes 21,561 trucks and 41,169 trailers but excludes TCVs of less than 3.5 tonnes. Therefore, given the fleet composition presented in Figure 3 and the fleet size of trucks and trailers (62,730 units of trucks and trailers) from Cold Chain Federation, we estimated temperature-controlled vehicles in the UK may be around 132,931. This includes 65,017 vans, 21,561 trucks, 41,169 trailers and 5,184 e-cargo bikes.

This was obtained using the following approaches:

Assuming we have X numbers of TCVs in the UK and the numbers of trucks and trailers are as indicated by Cold Chain Federation. If C, V, M and N represent numbers of cargo bike, vans, trucks, and trailers respectively in the UK; then,

$C+V+M+N=X$ [1]

From CCF 2023

$M+N=62{,}730$ [2]; then equation [1] can be modified as in [3].

$C+V+62{,}730=X$ [3]

Using our survey's percentage composition in Figure 3; ($C = 0.039X$ and $V = 0.4891X$).

$0.039X+0.4891X+62{,}730=X$ [4]

From equation 4, X is calculated to be 132,931.

However, the estimated 132,931 TCVs in the UK is considered a conservative figure due to the following reasons:

- CCF figure only accounts for its 278 members and there are temperature-controlled businesses in the UK that are not CCF members.
- There was 137.8 billion tonne kilometres cabotage (goods moved by a transport operator from another country) in the UK in 2020. It is possible that significant numbers of these vehicles are TCVs given that e.g. the UK imports 84% of fruits and 47% of vegetables. Hence, the actual numbers of TCV on the UK roads may be above our estimates [1].
- Logistics UK says their road members operate more than half of the UK's HGV fleet and more than one million vans and it is likely a substantial part of this is refrigerated vehicles.

Therefore, the survey result and other references found in the literature are used to obtain range for the refrigerated vehicles in the UK (Table 2) as detailed in Appendix I.

Table 2: Estimates of refrigerated vehicles in the UK

Vehicle type	Range
Vans	44,877 - 81,802
Trucks	21,561 - 39,301
Trailers	41,778 - 76,152
E-cargo	4,391 – 8,004
Total	112,608 – 205,260

Our analysis indicates penetration of e-cargo bikes on cold chain fleet composition, particularly for last-mile food delivery. This may be due to the known economic and environmental benefits of e-cargo bikes. Cargo bikes make journeys faster and more efficient, plus delivering improved air quality, which contributes to healthier and safer streets [17]. Leonardi et al. [18] assessed the impact of replacing diesel vans with electrical vans and cargo bikes in London and found a reduction of 55% in emitted CO_2 per parcel. The study also indicated a 20% decrease in total distance travelled because of the delivery system. Due to increased interest, a study for Transport for London estimates cargo bikes could replace up to 17 per cent of van kilometres in central London by 2030 [17].

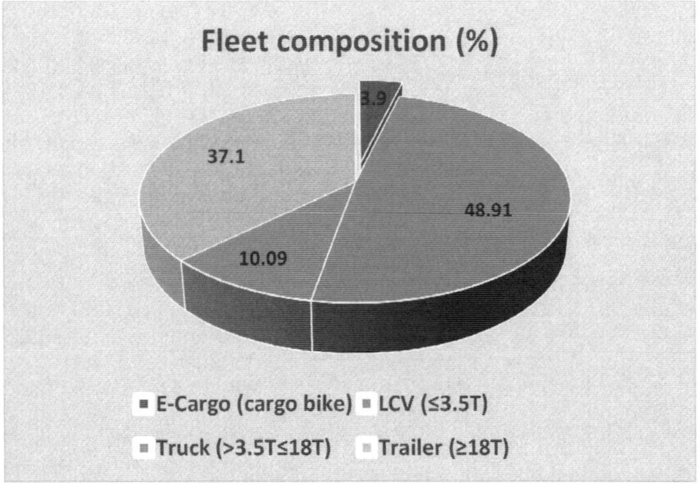

Figure 3: TRU percentage fleet composition

3.3 TRU Energy and Emissions

Energy consumption and emissions analysis are carried out using equations presented in Appendix II and as detailed in the references [16,17]. Data for modelling are as detailed in Table 3.

Table 3: Data input for TRU energy and emissions modelling.

Particular	Vans	Trucks	Trailer	Notes
Numbers of TRUs in the UK	44,877 – 81,802	21,561-39,301	41,778 -76,152	• Numbers of vehicles as estimated from survey and •Numbers of vehicles as estimated from survey, CCF data and literature. • •Details (capacity, refrigerant charge, leakage, etc.) of battery-driven compressor refrigeration units of the 5,184 E-cargo bikes are currently unknown. Thus, only energy-related emissions associated with E-cargo bikes are considered, while refrigerant emissions are not captured. Besides, considering the current numbers, the impact may be regarded as negligible. • The UK nationwide zero-emission vehicle accounts for 2.1% of all licensed vehicles in 2023[19]. Hence, about 1,365 vans representing 2.1% are modelled as electric vehicles with average energy consumption of 25.7 kWh/100km which is the average range for most delivery electric vans.
Annual TRUs running hours (hr/year)	2,167	2,299	2,299	• Results taken from survey responses are the annual running time of TRUs and not annual vehicle running hours.

Fuel consumption to run the refrigeration system (l/hr)	1.5	2.3	3.5	• Average TRU fuel consumption for trucks and trailers as indicated in literature [14][20][21] [15,16,22] and provided by a consultant managing over 10,000 vehicles in the UK. • Average TRU fuel consumption for vans as indicated in recent publication [20] where TRU fuel consumptions is 14.5% [14]. • Average energy consumption of cargo bikes is 55 miles/kWh, with a Transport for London report indicating an average distance of 90 km/rider/day [23].
Refrigerants	R404A, R452A R410a R134a R1270 + R744	R404A, R452A R410a R134a R1270 +R744	R404A, R452A R410a R134a R1270 + R744	• R404 is the predominantly used refrigerant in TRUs, accounting for between 80-90% [14,24]. • From the survey response, we estimate that the refrigerant usage may be around 83% for R404A, 14.81% for R134A, 1.79% for R452A, 0.36% for R410a and 0.04% natural refrigerant (R1270 + R744). • The GWP are 3922 (R404A), 2144 (R452A), 2088 (R410a),1430 (R134a), 2 (R1290) and 1 (R744) [25]. Based on the above percentage composition (refrigerant used) and GWP above, **3512.9** is the estimated average GWP and adopted for the refrigerant emissions evaluation.
Refrigerant Leakage	30%	11.4%	11.4%	• Refrigerant leakages of trucks and trailers are taken from the HFC Outlook Model [13]. • The refrigerant leakage rate of vans is taken from Francis et al. [11], which is empirical data of 327 delivery vans.
Refrigerant charge (kg)	2.0	2.5	5	Representative charge on most used TRUs as specified in literature[14,15,21,22,26].
Conversion factors	The conversion factor used for diesel is 2.63 kg CO_2e per litre of diesel and electricity 0.207 kgCO_2e/kWh.			

Depending on the number of vehicles, the annual energy consumption is as indicated in Table 4. The total energy consumption is estimated to be between 5.13 - 9.36 TWh. Previous studies by Foster et al. [12] and SKM [27] suggested indirect emissions from TRU between 1.02 and 1.20 MtCO_2e. Thus, putting TRU energy consumption within the UK between 4.64 TWh and 5.45 TWh using a carbon intensity of the grid of 0.22 kgCO_2e/kWh. However, this study was based on assumed 70,000 TRUs in the UK. As indicated in Table 4, trailers are the highest energy consumers, contributing around 71% to the UK's TRU energy demand while contributions from vans to energy use is around 4.5% despite its larger percentage of the total UK fleet.

Table 4: Annual energy consumption of TRUs by vehicle category.

Energy consumption	Energy consumption (TWh)
Vans	0.23 - 0.41
Trucks	1.24 - 2.27
Trailers	3.66 - 6.68
E-Cargo bike	0.0014 - 0.0025
UK Total	5.13 - 9.36

Our survey indicated that temperature-controlled fleet operators are increasingly purchasing lower/zero emissions vans and about 30% of their vans purchase in 2024 will be low/zero emissions vans. However, only 2% of their predicted trailer purchase for 2024 will be low emissions. This may be because of challenges associated with low emissions Heavy Goods Vehicles within the UK. This includes high cost of ownership, high maintenance cost and lack of required infrastructure such as charging stations. The chief executive officers of three different temperature-controlled businesses stated the following:

- *"Over 95% of our sales this year are traditional engine driven systems. Our 12-volt and 400-volt systems are available and generating a lot of interest, the sales of commercial electric chassis are not coming through while grid infrastructure are not sufficient to meet the depot plug requirements of large fleets with a lot of vehicles".*
- *"We all want a better future, but it's very hard to currently justify and test equipment without some good support, I have 2 trailers on demo with Greenchill fridges which are battery powered but the cost is huge and there is no support from anyone, which is pushing me to consider selling them".*
- *"The initial cost of anything in relation to alternative fuels is a large investment in comparison with the current diesel option, with drivers' wages, cost of fuel, cost of equipment which have increased by over 20% in the last 18-24 months and high interest rates. I think there would need to be some radical changes and support from the government to decarbonise, the hauliers are already pushed to the limits with rising costs and little understanding from their customers in particular the supermarkets. I think there would need to be tax benefits, grants and many other initiatives to convince the market to change, there is a will from most to be green and look green to their customers and the consumer, but they cannot foot all the costs. If anything, I feel we will become less green because, due to current strains, people are running older equipment much longer, which is less efficient, increasing pollution".*

Therefore, more must be done to support the industry and meet the UK emissions reduction targets. The emissions associated with TRUs in the UK are presented in Table 5.

Table 5: Analysis of emissions by source and vehicle category

	Scope 1		Scope 2
	Fuel GHG Emissions ($MtCO_2e$/year)	**Refrigerants GHG Emissions ($MtCO_2e$/year)**	**Electricity usage GHG Emissions (tCO_2e/year)**
E-cargo Bike	-	-	286.7 - 522.7
Vans	0.05 - 0.10	0.10 - 0.17	159.4 - 290.6
Trucks	0.30 - 0.55	0.02 - 0.04	-
Trailers	0.88 -1.61	0.08 - 0.15	-
UK Total	**1.23 - 2.26**	**0.20 - 0.36**	**446.1- 813.3**
	1.43 - 2.62		
	1.44 - 2.62		

Fuel used to power TRU contributes 86.1% of the emissions, while refrigerants-related emissions account for 13.9% of the total emissions, and 0.03% is contributed by electricity use. A previous study from Norwegian seafood put [28] fuel-related emissions between 63% and 87% while fuel use only accounts for around 62% of transport refrigeration emissions from another study [12].

4. Conclusion

Temperature-controlled vehicles in the UK are estimated to be around 158,934 ± 46,326; with e-cargo bikes becoming prominent in last-mile food delivery. The energy demand and emissions associated with transport refrigeration in the UK are estimated to be 5.13 - 9.36TWh and 1.44 -2.62 $MtCO_2e$/year. An estimated 71% of TRU energy consumption is linked to trailers. Still, while uptake of low-emissions vans is increasing, the trailer is foot-dragging mainly due to challenges related to infrastructure and total costs of ownership. There is great uncertainty in the number of vehicles and therefore on respective emissions. Therefore, it is recommended that ATP registration should be made compulsory for all TRUs so that it can serve as database of refrigerated vehicles in the UK.

References

[1] Evans J, Peters T. Towards a net-zero food cold-chain. IOR Annual Conference 2021 – The Journey to Net Zero Heating and Cooling – Beyond Refrigeration, Institute of Refrigeration; 2021.

[2] Sovacool BK, Bazilian M, Griffiths S, Kim J, Foley A, Rooney D. Decarbonizing the food and beverages industry: A critical and systematic review of developments, sociotechnical systems and policy options. Renewable and Sustainable Energy Reviews 2021;143:110856. https://doi.org/10.1016/j.rser.2021.110856.

[3] Li M, Jia N, Lenzen M, Malik A, Wei L, Jin Y, et al. Global food-miles account for nearly 20% of total food-systems emissions. Nat Food 2022;3:445–53. https://doi.org/10.1038/s43016-022-00531-w.

[4] Food and Drink Federation. Roadmap to Net Zero Overview for the UK food and drink sector. London: 2022.

[5] Hasnain S, Ingram J, Zurek M. MAPPING THE UK FOOD SYSTEM A Report for the UKRI Transforming UK Food Systems Programme ACKNOWLEDGEMENTS. n.d.

[6] Barb Renner, Brian Baker, Justin Cook, Jagadish Upadhyaya. Fresh vs. frozen-The future of fresh in a changing competitive landscape. USA: 2021.

[7] James Davey. Britons shift from fresh to frozen in cost-of-living crunch. Reuters 2022.

[8] BFFF. The Frozen Food Revolution is Here. British Frozen Food Federation 2023.

[9] Cold Chain Federation. COLD CHAIN REPORT. 2023.

[10] Tassou SA, De-Lille G, Ge YT. Food transport refrigeration – Approaches to reduce energy consumption and environmental impacts of road transport. Appl Therm Eng 2009;29:1467–77. https://doi.org/10.1016/J.APPLTHERMALENG.2008.06.027.

[11] CENEX. Refrigerated Transport Insights. Loughborough: 2021.

[12] Alan F, Tim B, Judith E. Carbon emissions from refrigeration used in the UK food industry. International Journal of Refrigeration 2023. https://doi.org/10.1016/J.IJREFRIG.2023.01.022.

[13] EPEE (European Partnerships for Energy and Environment). HFC Outlook Model. European Union, 2023.

[14] Francis C, Davies G, Graeme Maidment, Judith Evans, Edward Hammond, Andrew Gigiel. Sustainable Refrigerated Road Transport-Investigating the Scale of Carbon Emissions from Direct-Drive Last Mile Refrigerated Vehicles. The knowledge hub for refrigeration, London: 2019.

[15] Rai A, Tassou SA. Environmental impacts of vapour compression and cryogenic transport refrigeration technologies for temperature controlled food distribution. Energy Convers Manag 2017;150:914–23. https://doi.org/10.1016/j.enconman.2017.05.024.

[16] Tassou SA, Kolokotroni M, Gowreesunker B, Stojceska V, Azapagic A, Fryer P, et al. Energy demand and reduction opportunities in the UK food chain. Proceedings of the Institution of Civil Engineers - Energy 2014;167:162–70. https://doi.org/10.1680/ener.14.00014.

[17] Dybdalen Å, Ryeng EO. Understanding how to ensure efficient operation of cargo bikes on winter roads. Research in Transportation Business & Management 2022;44:100652. https://doi.org/10.1016/j.rtbm.2021.100652.

[18] Leonardi J, Browne M, Allen J. Before-After Assessment of a Logistics Trial with Clean Urban Freight Vehicles: A Case Study in London. Procedia Soc Behav Sci 2012;39:146–57. https://doi.org/10.1016/j.sbspro.2012.03.097.

[19] Department for Transport. Vehicle licensing statistics: April to June 2023. National Statistics 2023.

[20] Fabris F, Artuso P, Marinetti S, Minetto S, Rossetti A. Cooling unit impact on energy and emissions of a refrigerated light truck. Appl Therm Eng 2022;216. https://doi.org/10.1016/j.applthermaleng.2022.119132.

[21] Cenex Advanced Technology Innovation Centre. Auxiliary temperature reduction units in the Greater London area. London: 2018.

[22] Rai A. Energy demand and environmental impacts of food transport refrigeration and energy reduction methods during temperature-controlled distribution. PhD. Brunel University London, 2019.

[23] Transport for London. Cargo bike action plan. London: 2023.

[24] Cavalier G, Devin E, Michineau T, Vannson F, Thomas G. Refrigerant Leakage in the Refrigerated Transport Sector. 4th IIR Conference on Sustainability and the Cold Chain, Auckland: International Institute of Refrigeration (IIR),; 2016.

[25] International Institute of Refrigeration. Global warming potential (GWP) of HFC refrigerants. International Institute of Refrigeration 2022.

[26] Zemo Partnership. Emissions Testing of Two Auxiliary Transport Refrigeration Units. London: 2021.

[27] Enviros SKM. Use of Refrigeration in UK Soft Drinks Supply Chain: Sinclair Knight Merz (Europe) Ltd 2014.

[28] Ziegler F, Winther U, Hognes ES, Emanuelsson A, Sund V, Ellingsen H. The Carbon Footprint of Norwegian Seafood Products on the Global Seafood Market. J Ind Ecol 2013;17:103–16. https://doi.org/10.1111/J.1530-9290.2012.00485.X.

Appendix I: Numbers of Refrigerated Vehicles in the UK

There are various figures reported for temperature-controlled vehicles in the UK. The figures are presented (Table 6) on numbers of refrigerated vehicles in the UK.

Table 6: Various references on the number of refrigerated vehicles in the UK

SN	Year	Numbers reported	References
1	2009	140,000	UK refrigerated road transport market. (iifiir.org)
2	2013	172,000	Special_Report_UK_Distribution.pdf (coldchainnews.com)
3	2014	84,000 growing at 10% per annum.	AIR0043 - Evidence on Action on air quality (parliament.uk)
4	2015	84,000 (vans - 50,000, trucks -16,000, trailers -18,000)	Liquid-Air-on-the-European-Highway.pdf (airqualitynews.com)
5	2021	100,000	Greening Refrigerated Transport - Financial and environmental impacts (cenex.co.uk)
6	Current calculation	132,931	CCF data + survey

The above references are normalised to 2023 using 25 years (1997 - 2022) average annual growing rate of LCV and HGV in the UK as detailed by the Department of Transport (Vehicle licensing statistics: 2022 - GOV.UK (www.gov.uk)). During this period, LCV grows at an annual rate of 4.04% while HGV grew at 0.72% per annum. The average is then used to obtain values in Table 7.

Table 7: Normalised estimates of refrigerated vehicles in the UK.

SN	Reference year	Ref year value	2023 Estimates
1	2009	140,000	195,132
2	2013	172,000	218,035
3	2014	84,000	198,000*
4	2015	84,000	104,649
2	2021	100,000	104,858
6	2023	This work	132,931

*reference mentioned 10% annual growth rate.

The 2023 estimates are then used to obtain a range for the temperature-controlled vehicles in the UK which is thought to be 158,934 ± 46,326, this range is based on the average and standard deviation presented in table 7. Of all references above, only reference 3 detailed vans, trucks and trailers as 50,000, 16,000 and 18,000 respectively. However, the 2023 estimates for trucks and trailers from the surveys carried out is 36,008 which is lower than 62,730 (trucks and trailers) reported by CCF which is a trade association representing around 278 temperature-controlled businesses in the UK. The proportion the UK temperature-controlled businesses who are members of CCF is unknown and it is likely that the numbers of refrigerated trucks and trailers in the UK is more than what is reported by the CCF. For instance, Logistics UK says their members operate more than half of the UK's HGV fleet and more than one million vans and it is likely a substantial part of this is refrigerated vehicles, however exact figures are unknown. Therefore, the estimated 158,934 ± 46,326 refrigerated vehicle is adopted, and percentage composition obtained from survey is used to obtain numbers of vans, trucks, and trailers. Hence, the breakdown on vehicle category is presented in Table 8.

Table 8: Estimates of refrigerated vehicles in the UK

Vehicle type	Range
Vans	44,877 - 81,802
Trucks	21,561- 39,301
Trailers	41,778 - 76,152
E-cargo	4,391 – 8,004
All	112,608 – 205,260

Appendix II

Annual total Emissions= Scope 1 (Fuel and Refrigerant Emissions)+

Scope 2 (Electricity Related Emissions) (5)

Annual fuel Emissions= $(AFC \times EM_{fuel})$ (6)

Annual refrigerant Emissions= $(C \times GWP \times ALR)$ (7)

Annual electricity Related Emissions= $(AEC \times EM_{electricity})$

Where:

AFC=Annual Fuel Consumption (L)

AEC=Annual Electricity Consumption (kWh)

$EM_{electricity}$ *=Electricity Emission factor* $\left(\frac{kgCO_{2\text{-}eq}}{kWh}\right)$

EM_{fuel} *=Fuel Emission factor* $\left(\frac{KgCO_{2\text{-}eq}}{L}\right)$

C=Refrigerant Charge (kg)

GWP=Global Warming Potential $\left(\frac{KgCO_{2\text{-}eq}}{Kg\ refrigerant}\right)$

ALR=Annual Leakage Rate (% of norminal initial charge)

About the authors

Dr Rasaq Lamidi

Rasaq Lamidi has a PhD in Energy (Renewable energy Systems) from Newcastle University, MSc in Carbon and Resource Management from University of Central Lancashire, Preston, the UK and BSc in Food Science and Technology from University of Agriculture, Abeokuta, Nigeria. He is currently working as Research Fellow at the Centre for Sustainable Cooling, School of Chemical Engineering, University of Birmingham on various projects focusing on sustainable cooling and cold-chains.

Dr Xinfang Wang

Dr Xinfang Wang is an Assistant Professor and the Group Leader of Resilient Systems for Energy and Cooling at the Birmingham Energy Institute in University of Birmingham. She has led several multi-million projects in the UK and EU on food cold chain, including Zero Emission Cold Chain (ZECC) and European food chain supply to reduce GHG emissions by 2050 (ENOUGH) . She also sits on the Research Committee for the UK Energy Research Centre (UKERC) and advises on the research strategy and priorities. She is the Equality, Diversity and Inclusion (EDI) Champion for UKERC.

Q & A Emissions Assessment of Refrigerated Food Transport in the UK

Daniel Colbourne - A couple of questions about the survey response. Is the fleet composition percentage of vehicles or mass of product?

Xinfang Wang - They are the fleet operations.

Ken Gleeson - The survey says the number of respondents from TRU OEMs is 19%. Can you provide the actual number?

Xinfang Wang - Not off the top of my head but we received about 26 responses out of the 139 stakeholders.

Tom Southall - Some useful and interesting data. The title of the paper refers to 'food road transport' but I think your figures include all transport refrigeration (which includes pharma, life sciences and plants)? It is good if this is the case as you have then covered the whole industry, rather than a subset.

Xinfang Wang - Most of the respondents were supermarkets involved with food transportation but we also had responses that included pharmaceutical transport.

Tom Southall - Can you explain how an e-cargo bike uses electricity for refrigeration. Are these not just passive cool bags.

Xinfang Wang - I'll have to come back to you after the webinar.

Gabrielle Hoffman – How much refrigerant is in the vans, trucks and trailers (total filling amount)?

Xinfang Wang - In the paper we have specified how we calculated the emissions from the refrigerants and we have used a weighted average of the different types of refrigerant that were used and we have used some assumptions for refrigerant leak and recharge rate.

Norman Hignam – The average charge is between 1 and 1.5kg

Tom Thurnham - Could you clarify whether the energy consumption estimates are for the refrigeration and vehicle engine, or just the onboard refrigeration unit?

Xinfang Wang - They are just for the TRU, refrigeration units.

Daniel Colbourne - What is the % split of emissions between energy-related and refrigerants? How does this split compare when using 20 y ITH GWP instead of 100 y ITH GWP?

Xinfang Wang - Refer to slide Results – TRU emissions. For vans most of the emissions come from refrigerant emissions and for trucks and trailers most of the emissions come from fuel use to power the transport refrigeration unit.

Ben Fielden -You mentioned that non-diesel powered trailers represent only 1.74% of vehicles purchased in 2022. Is there a further breakdown of what these non-diesel solutions are?

Xinfang Wang - We did not ask for this information specifically but most non-diesel options are electric alternatives.
Chris Rhodes - I see R404A was used for calculations. All large manufacturers changed to R452A in 2016/2017 which is now the vast majority as they stopped selling new systems with R404A. Why was R404A used for the calculations?

Xinfang Wang - That is based on the survey responses that we received. We asked stakeholders what they mainly used and they gave us the information. It may be that others who haven't responded use other refrigerants but we had to base our calculations on the survey response. If we could get more stakeholders to provide us with information anonymously for our calculations then that would be really helpful. If other stakeholders feel they have other information then please get in touch and we would be happy to revise our results and calculations if we get more data.

Norman Highnam - Now the report has been published how does the DFT look to address the emissions?

Xinfang Wang - We have provided our report and feedback to the DFT and other projects funded by the DFT and provided the list of suggestions for policy and decision makers to consider. i.e. The use of ATP to become compulsory to track the number of TRUs

Daniel Colbourne - How about so-called "cryogenic" systems, such as N2 or CO2 direct injection - how do the emissions of these systems compare against conventional diesel/compression systems?

Xinfang Wang - From the information we gathered from the stakeholders that is not an option that has been considered. We need more stakeholders to respond and more data.

Daniel Colbourne - Will you be making/have you made future projections of emissions? Is there an expectation on growth in electric vehicles and how does that impact on this split between energy and refrigerant emissions?

Xinfang Wang - We are making predictions of the emissions and energy use up to 2050 and we are developing different scenarios for the predictions according to different factors, technology, finance policy and consumer behaviour change. We will share our findings for those predictions in the near future. We are looking at both areas of the split in terms of decarbonising, for refrigerant transferring to natural refrigerants or low carbon refrigerants and emissions from fuel use we are looking at electric alternatives and also looking at a change of logistics related to the composition of fleets such as vans and large cardboard bags because of the change to more frequent online shopping instead of large weekly deliveries. We are continuing research in those areas and developing benchmarking of refrigerated transport in the UK, again, getting data is the challenge but we are using what we have to give stakeholders information.

Ken Gleeson - Trailer TRU fuel consumption. Did you assume 3.5L/hr for each hour of use ? If so, this will greatly over state the actual fuel consumption. TRU fuel consumption will vary greatly depending on a number of factors but especially in terms of fresh vs frozen set-point and whether at set-point or pulling down to set-point.

Xinfang Wang - For our model we used an average for different types of vehicle but we agree there are large differences among the different types of vehicles and that can lead to some uncertainty in our analysis. We could do a further sensitivity analysis because of this.

Ben Fielden - Was this evidence submitted in the recent Non-Road Mobile Machinery Decarbonisation Options call for evidence by the Dept for Transport?

Xinfang Wang - I hope so but I can't speak for the government department.

Gabrielle Hoffman - Is the study available officially and where?

Xinfang Wang - The study will be available at the beginning of next year. We have done the analysis of energy and emissions and we will be making predictions up to 2050 and developing benchmarking. We have this paper available through the IOR and will continue making further papers available when they are ready.

Norman Highnam - Do we need to be looking more at the use of non-refrigeration systems and focus more on the need to insulate rather than refrigerate in the transport sector in the UK?

Xinfang Wang - Both. We need to try to reduce demand, to try to improve efficiency and reduce demand on energy and we also need to look for low emission alternatives too. Both options are essential.